国家林业和草原局普通高等教育"十四五"规划教材

普通植物病理学实践教程(双语)

Practice Tutorials for General Phytopathology

李培琴 主编

中国林业出版社
CFPH China Forestry Publishing House

内 容 简 介

教材介绍了普通植物病理学实践教学的基本内容，包括25个实验和3个教学实习，除了包含传统的观察性和验证性实验外，还加入若干综合性和设计性实验，旨在通过野外植物病害调查、标本采集和病原物鉴定等实践操作，提升学生对普通植物病理学知识的掌握。

本教材既可以作为植物保护、森林保护本科专业普通植物病理学课程的实践课指导教材，也可作为林学、园林等专业林木病理学和园林植物保护课程的实践教学指导教材，还可作为植物病理学研究人员的工作参考书。

图书在版编目(CIP)数据

普通植物病理学实践教程：汉文、英文/李培琴主编.—北京：中国林业出版社，2022.12(2023.10重印)

国家林业和草原局普通高等教育“十四五”规划教材

ISBN 978-7-5219-2016-1

Ⅰ.①普… Ⅱ.①李… Ⅲ.①植物病理学-教材-汉、英 Ⅳ.①S432.1

中国版本图书馆CIP数据核字(2022)第241575号

策划编辑：范立鹏
责任编辑：范立鹏
责任校对：苏 梅
封面设计：周周设计局

出版发行：中国林业出版社
(100009，北京市西城区刘海胡同7号，电话010-83223120)
电子邮箱：cfphzbs@163.com
网址：www.forestry.gov.cn/lycb.html
印刷：北京中科印刷有限公司
版次：2022年12月第1版
印次：2023年10月第2次
开本：787mm×1092mm 1/16
印张：13.375
字数：317千字 数字资源：44个
定价：46.00元

《普通植物病理学实践教程》(双语)
编写人员

主　　编: 李培琴

副 主 编: 张丹凤

编写人员: (按姓氏笔画排序)

白露超(青海大学)

闫金姣(西北农林科技大学)

李培琴(西北农林科技大学)

张丹凤(合肥工业大学)

单体江(华南农业大学)

费昭雪(西北农林科技大学)

徐利剑(黑龙江大学)

前 言
Preface

实践教学是高等院校最基本的教学形式之一。实践教学可以培养和提高学生的自主学习能力、创新意识与创新能力，促进学生动手动脑，对培养学生科学思维和实践创新精神，全面推进素质教育有着重要的作用。“普通植物病理学”为植物保护、森林保护等本科专业的专业基础课，其实践环节不仅能巩固学生对理论知识的理解，还能提高学生的专业技能与创新意识。

随着经济和信息的全球化，如何培养既懂专业知识又精通外语的复合型人才，已成为教育机构和教育工作者亟待解决的重要问题。而实施双语教学是培养学生成为具有国际合作意识、国际交流与竞争能力的国际化人才的重要途径。

为了紧跟时代和学科的发展，培养具有创新思维和创新能力的卓越人才，配合双语教学的实施，我们根据教学大纲和课程学时要求，在立足编者多年教学和科研经验基础上，借鉴兄弟院校教学经验，参考英文原版教材和公开发表的相关论著，经过精心筛选和提炼，编写了此书。

本教材包括 25 个实验和 3 个教学实习，所选内容除了包含传统的观察性和验证性实验外，还加入若干综合性和设计性实验。特别是本教材还介绍了普通植物病理学实践教学的基本内容，旨在通过野外植物病害调查、标本采集和病原物鉴定等实践操作，进一步提升学生对本专业课程知识的认识和掌握。

本教材以理论为基础，重视理论结合实践，整体上体现了普通植物病理学实践教学的特色。同时，教材减少了观察性实验的比例，提高了操作技能训练的比例。通过观察和实践操作，旨在帮助学生理解课程的基础理论知识，掌握课程相关的基本技能，培养学生的观察、思考和动手能力。

本教材由李培琴担任主编，张丹凤担任副主编，各章节编写分工如下：第一部分由李培琴和费昭雪编写；第二部分的实验一、实验十、实验二十至实验二十五由张丹凤、李培琴、闫金娇和徐利剑编写；第二部分的实验二至实验五、实验九、实验十一至实验十九由李培琴编写；第二部分的实验六至实验八由单体江和李培琴编写；第三部分由李培琴、张丹凤、白露超和费昭雪编写；附录Ⅰ、Ⅱ、Ⅳ和Ⅴ由李培琴编写；附录Ⅲ由张丹凤和闫金娇编写；教材插图由李培琴制作，全书最后由李培琴统稿定稿。

本教材既可以作为植物保护、森林保护本科专业普通植物病理学课程的实践课指导教材，也可作为林学、园林等专业林木病理学和园林植物保护课程的实践教学指导教材，还可作为植物病理学研究人员的工作参考书。

教材建设是一项长期的工作。限于人力及时间，在实验内容编排和中英文表述方面尚存欠妥之处，恳请广大读者提出宝贵意见和建议，以便修订和再版。

编 者

2022 年 9 月

目　录

Contents

第一部分　植物病理学实验室守则
PART ONE　Rules for Plant Pathology Laboratory

Ⅰ. 实验室基本要求

1. 参加实验的学生，要遵守学习纪律，按时进入实验室，在指定的实验位置就座，不得无故缺席，按时完成实验任务。

2. 每次实验前要充分预习实验指导，明确实验的目的和要求、原理和方法、内容和作业等，以保证实验顺利进行。

3. 每次实验前，必须做好实验用具的检查工作。

4. 实验进行中，严格遵守课堂秩序，不得高声交谈和随意走动，有疑问直接请教老师。实验操作要小心谨慎，认真观察实验现象，做好实验记录。

5. 实验结果记录和绘图应实事求是，不准任意改动，互相抄袭。实验报告要用统一的实验报告纸和铅笔完成，要求字迹清楚，绘图规范，按时完成实验报告。

6. 爱护实验仪器，节约实验试剂。

7. 注意实验室安全，使用易燃易爆、有毒有害试剂时要按使用规程及要求操作，禁止用酒精灯互相点火。

8. 实验完毕应将仪器放回原位，整理好实验用具，值日生将实验室打扫干净，关好门窗和水电，经老师检查确定后方可离开。

Ⅰ. Basic Laboratory Requirements

1. Students participating in the experiments must abide by the learning disciplines, enter the laboratory on time, and sit on the designated experimental seats. Absence is not allowed if there is no legitimate reason. Please complete the experiment on time.

2. Please preview the experimental guidance before each experiment, and clarify the purposes, requirements, principles, methods, contents, and tasks of each experiment to make sure the experiment goes smoothly.

3. Check the experimental equipment before each experiment.

4. Please obey the classroom orders strictly when the experiment is going on. Please do not talk loudly or walk around. Please consult the teacher directly if there is any question. Be cautious in experiment operation, be careful to observe experimental phenomena and make

experimental records.

5. It should be truthfully to record experimental results and draw, and it is not allowed to change and copy any results. Please complete the experimental report on time using the unified experimental papers with clear handwriting and standard drawing.

6. Please take good care of laboratory equipment and save medicines and reagents.

7. Do pay attention to laboratory safety. Please accord to the manipulation procedures and operation requirements strictly when using flammable, explosive, toxic, and harmful drugs. Do not use alcohol lamps to ignite each other.

8. The instrument should be kept on position after the experiment is finished, and the laboratory apparatus should be tidied up. The students on duty should clean up the laboratory, close the doors, windows, water tap, and electricity, and leave the laboratory after the teacher's check and confirmation.

Ⅱ. 实验室安全知识

实验室安全知识学习要放在第一次进入实验室和第一次实验课前进行。由于实验室经常使用到水、电、高压设备、玻璃制品，以及具有毒性、挥发性、腐蚀性或爆炸性的化学药品，因此，实验人员必须重视安全工作。

1. 进入实验室开始工作前，应了解水阀、电阀、门窗所在的位置。

2. 使用酒精灯时必须注意安全。不得向点燃的酒精灯内添加酒精。添加酒精时不应超过酒精灯容积的2/3。熄灭时用灯帽盖灭，不能用嘴吹灭。万一洒出的酒精在桌子上燃烧起来，切勿慌乱，应立即用湿布扑灭。

3. 实验用过的菌种及带有活菌的各种器皿应先经高压灭菌后才能洗涤，特别对于检疫性实验材料，必须进行灭活处理，绝不能扩散出实验室。

4. 进行高压蒸汽灭菌时，应严格遵守操作规程，灭菌人员在灭菌过程中不得离开灭菌室。

5. 使用电器设备(如烘箱、培养箱、离心机、微波炉)时，应严防触电，绝不可用湿手触及电器开关。用电笔检查电器设备是否漏电，凡是漏电的仪器，一律不得使用。

6. 使用强腐蚀性溶液，如浓酸和浓碱，务必小心操作，防止迸溅。如不慎触及皮肤，应立即治疗。

7. 挥发性危险化学品的使用必须在通风橱内进行。

8. 使用可燃物特别是易燃物时，应严格遵守易燃物使用规定。

9. 废液处理回收时，应按规定将其存放在指定位置和相应容器内，集中处理。

10. 有毒物品应按实验室的规定在办理审批手续后领取，使用时严格操作，用后妥善处理。

Ⅱ. Laboratory Safety Knowledge

The learning of laboratory safety knowledge should be carried out before first entering the laboratory and the first experimental class. In the laboratory, water, electricity, high-preesue containers, high-temperature equipments, fragile glasswares, and some toxic, volatile, corrosive, flammable or explosive chemicals might be used. Therefore, it should be pay sufficient attention to experiment safety.

1. Please be aware of the location of the water valve, electric valve, doors and windows before working in the laboratory.

2. Please use the alcohol lamps securely. Do not add alcohol to a burning alcohol lamp and the volume of alcohol in the alcohol lamp should not exceed 2/3 of its volume. Extinguish the alcohol lamp with a lamp cover but not with the mouth. Please do not panic if the spilled alcohol burns on the table, which could be covered with a damp cloth immediately.

3. The pathogen cultures and the different utensils with live pathogens should be treated by high temperature and pressure. The quarantine experimental materials should be inactivated and it is forbidden to spread out of the laboratory.

4. Strictly follow the operating procedures to conduct autoclaving, and the operator mustn't leave the laboratory.

5. Take care when using electrical equipment(such as ovens, incubators, centrifuges, and microwave ovens) to avoid electric shock. Do not touch electrical switches with wet hands. Use an electric pen to check whether the electrical equipment is leaking, and do not use any leaking instrument.

6. Be extremely careful when using strong corrosive solutions, such as concentrated acid and concentrated alkali, to prevent splashing. An immediate medical treatment is necessary if skin contact happens.

7. It should be conducted in a fume hood when using volatile hazardous chemicals.

8. Follow the application regulations strictly when using combustible materials.

9. Store the waste liquid in a designated container at a designated location for centralized processing.

10. Toxic substances should be collected after going through the examination and approval procedures according to the laboratory administrations. Strictly use these chemicals, and dispose them properly after using.

Ⅲ. 实验室急救

在实验过程中若不慎发生伤害事故，应立即采取适当的急救措施。

1. 玻璃割伤：首先必须检查伤口内有无玻璃碎片，然后用硼酸溶液洗净，再涂擦碘酒，必要时用纱布包扎。若因伤口较大或过深而大量出血，应迅速在伤口上部和下部扎紧止血带，并立即到医院治疗。

2. 烫伤：一般用医用酒精消毒后，涂擦苦味酸软膏。若伤处出现红痛或红肿，可擦医用橄榄油或用无菌棉花蘸取医用酒精敷盖伤处；若皮肤起泡，不要弄破水泡，防止感染；若伤处皮肤呈棕色或黑色，应用干燥的无菌消毒纱布轻轻包扎好，尽快送医院治疗。

3. 烧伤：烧伤自行处理是很重要的，其紧急处理的5个步骤为“冲”“脱”“泡”“包”和“送”。“冲”是指烧伤后立即脱离热源，用流动的冷水冲洗伤面，降低创面温度。“脱”是指脱去衣服，正确的处理方法是边冲边脱。“泡”是指将伤口继续浸泡在冷水中。“包”是指包裹创面，送医院之前一定要包裹创面，裹上一块干净的毛巾也可，切忌随便涂抹药膏。“送”是指送医就诊，寻求医生的救助。

4. 灼伤：当强碱(如氢氧化钠、氢氧化钾)及金属钠、钾等碱性化学药品触及皮肤而引起灼伤时，要先用大量自来水冲洗，再用5%的硼酸溶液或2%的乙酸溶液涂洗。当强酸、溴、氯、磷或其他酸性化学药品触及皮肤而致灼伤时，应立即用大量自来水冲洗，再以5%的氢氧化铵溶液洗涤。如酚类试剂触及皮肤引起灼伤，可用医用酒精洗涤。

5. 触电：触电时可按下述方法之一切断电路：①关闭电源；②用干木棍使导线与接触者分开；③使触电者和地面分离。急救者必须采取防止触电的安全措施，手和脚必须绝缘。

Ⅲ. Laboratory First Aid

If an accident occurs during the experiment, appropriate first aid should be taken immediately.

1. Cut by glass: Check whether there are any glass fragments in the wound firstly, clean the wound with the boric acid solution, and then apply iodine. If necessary, wrap the wound with gauze. If the wound is large or deep and causes a lot of bleeding, blood vessels should be fastened on the upper and lower parts of the wound to stop bleeding, and go to the hospital for treatment immediately.

2. Burns: Generally, apply picric acid ointment after disinfecting with medical alcohol. If the wound is red, painful, or swollen, rub medical olive oil or use alcohol cotton to cover the wound. If the skin blisters, do not break it to prevent infection. If the wound is brown or black, wrap it gently using dry and sterile gauze, and go to the hospital for treatment as soon as possible.

3. Fire burns: It is very important to treat burns by themselves. There are five steps for emergency treatment: rinsing, taking off, soaking, wrapping and sending. Rinsing means using running cold water to wash the wound surface to decrease the temperature. Taking off means taking off the burning clothes, and the preferable way to handle it is to rinse it when taking off. Soaking means soaking the wound in cold water continuously. Wrapping means wrapping the wound before going to the hospital, such as wrapping it with a clean towel, and do not apply ointment casually. Sending means going to the hospital medical treatments.

4. Ambustion: For the ambustion caused by strong alkalis (such as sodium hydroxide, potassium

hydroxide), metal sodium, potassium, and other alkaline chemicals, rinse it with a large amount of tap water immediately, and then wash it using 5% boric acid solution or 2% acetic acid solution. For the ambustion caused by strong acid, bromine, chlorine, phosphorus, or other acidic chemicals, wash it immediately using a large amount of tap water and then wash it using 5% ammonium hydroxide solution. If the ambustion is caused by phenolic reagents, wash it using medical alcohol.

5. Electric shock: Cut off the circuit according to one of the following methods after an electric shock: ①Turn off the power; ②Separate the wire from the contact using a dry wooden stick; ③Separate the contact from the ground. First-aiders must take safety measures to prevent electric shock and their hands and feet must be insulated.

第二部分　普通植物病理学实验
PART TWO　Experiments of General Plant Pathology

实验一　植物病害症状的观察

【概述】

植物病害是植物与病原在外界环境条件影响下相互作用并导致植物生病的过程，可分为非侵染性病害和侵染性病害。非侵染性病害是指由不良的环境因子(如水分失调、光照失调、营养失调、温度失调、大气污染、农药药害等)所引起的植物病害；侵染性病害是指由生物因素(主要包括菌物、细菌、病毒、线虫、寄生性植物等)引起的植物病害。

患病植物由于异常生理活动而在受病组织内部或外部会表现异常状态，即植物症状，包括形状、质地、颜色、气味和存在的病原物结构等。

各种植物病害的症状具有相对的特殊性和稳定性，因此，症状是诊断植物病害的重要依据之一。要熟练地利用症状来识别病害，必须具备植物病害基础知识和一定的实践积累，最好的方法是进行野外实地观察。

植物病害症状可分为病状和病征。病状是指植物患病后其本身所表现的不正常状态，如变色、坏死、腐烂、萎蔫和畸形；病征则是指病原物在患病部位所表现的特殊结构，如粉、霉、点、蕈菌、菌脓和流胶等。

【实验目的】

1. 观察认识植物病害的主要病状、病征类型。
2. 理解植物病害各类症状术语的含义。

【材料和器具】

1. 实验材料

实验室保存的各种类型的植物病害标本，如杨树锈病、花椒落叶病、正木白粉病、柳树白粉病、葡萄霜霉病、花椒叶锈病、正木炭疽病、杨树灰斑病、苹果黑星病、梭梭白粉病、臭椿白粉病、核桃白粉病、杨树煤污病、杨树腐烂病、杨树溃疡病、杨树黑斑病、葡萄毛毡病、板栗疫病、苹果花叶病、芍药叶霉病、核桃细菌性黑斑病、松苗猝倒病、桃缩叶病、杨树缩叶病、泡桐丛枝病、松瘤锈病、樱花冠瘿病、毛白杨根癌病、国槐枝枯病、花椒干腐病、花椒根腐病、花椒根结线虫病等。

2. 实验器具

计算机、投影仪、显微镜、手持放大镜、解剖针、载玻片、盖玻片、镊子、剪刀等。

【方法和步骤】

1. 根据附录Ⅰ的描述，详细观察各类植物病害的症状。

2. 列表记录供试标本的发病部位、症状类型、有无病征及其类型。

【结果和讨论】

1. 根据本实验提供的植物病害标本填写表2-1。

表2-1 植物病害症状观察记录表

寄主植物	病害名称	发病部位	病状特点	病征特点	病原种类

2. 病状和病征对植物病害诊断有什么作用?

3. 如何区分侵染性与非侵染性植物病害?

EXPERIMENT 1 Observation of Plant Diseases Symptoms

【Introduction】

Plant disease refers to the process that the physiological functions of plant is disordered and its structure tissues are destroyed when the plant is infected by pathogen or influenced by external improper environment condition, which can be divided into infectious disease and noninfectious disease. The plant diseases caused by biotic factors, such as fungi, bacteria, virus, nematode and parasitic plants, are called as infectious disease. Noninfectious plant diseases are caused by adverse environmental factors, such as water imbalance, illumination imbalance, nutrition disorder, temperature disorder, air pollution and pesticides phytotoxicity, etc.

The diseased tissues usually exhibit abnormal phenomena inside or outside due to the disorder of physiological activities of diseased plants, which is also called symptom.

The symptom of each plant disease shows a certain particularity and stability, which is the important basis for the diagnosis of plant diseases. To skillfully use symptoms to identify diseases, it is necessary to have the basic knowledge of plant diseases and a certain accumulation of practice, and the best way is to conduct field.

Symptom can be divided into morbidity and sign. Morbidity refers to the abnormal state of the plant itself after it is sick, such as discoloration, necrosis, rot, wilt and malformation, while sign refers to the special structures of pathogens in or on the diseased tissues, such as powder, mildew, particles, mushroom, ooze and gummosis, etc.

【Experimental Purpose】

1. Observe and recognize the main types of morbidity and sign.

2. Understand the terms for the symptoms of plant diseases.

【Materials and Apparatus】

1. Materials

Different types of plant disease specimens preserved in laboratories, such as poplar rust, gray spot, sooty mold, rot, fern leaf, canker, black spot and root tumor; prickly ash leaf cast, leaf rust, stem canker, root rot and nematode root knot; grape downy mildew and gray mold; apple scrab, mosaic, ring rot and stem canker; walnut anthracnose, bacterial black spot, powdery mildew and branch blight; pine needle cast, stem gall rust, nematode wilt and seedling damping off; peach fern leaf and soft rot; phyllody of rose and peony; witch's broom of bamboo and paulownia; wheat rust, powdery mildew, scab and dwarf smut; rice blast and false smut; corn smut, northern leaf blight and southern leaf bight, ect.

2. Instruments and Appliances

Computers, projectors, microscopes, handheld magnifying glasses, dissecting needles, slides, cover slides, tweezers, scissors, etc.

【Methods and Procedures】

1. Observe the symptoms of various plant diseases in detail according to the description in the Appendix I.

2. Record the diseased tissue, symptom type, sign and its type of plant disease specimens provided in this experiment.

【Results and Discussion】

1. Fill in Table 2-1 according to the plant disease specimens provided in this experiment.

2. Discuss the functions of morbidity and sign for the diagnosis of plant disease.

3. How to distinguish infectious and noninfectious plant diseases?

Table 2-1 Record Table for Symptom Observation of Plant Disease

Host plant	Plant disease	Disease tissue	Morbidity	Sign	Pathogen type

实验二 卵菌重要植物病原物的识别与鉴定

【概述】

卵菌是重要的植物病原物之一。其中，霜霉目为卵菌中最重要的植物致病类群，可引起植物的霜霉病、根腐病、疫病、猝倒病、种子霉烂病及白锈病等。

卵菌的习性由腐生到专性寄生、由水生向陆生变化，与之对应，在繁殖体形态上表现出不同的特征，如孢囊梗由菌丝状到显著分化，孢子囊由不脱落到脱落(图 2-1)；传播媒介也相应地由水流传播过渡到气流传播；寄生部位由根茎交接的土壤表面到地上的茎叶部；所致病害的症状由腐烂、坏死到促进性病状等。

卵菌的营养体为发达的无隔菌丝体。无性繁殖产生游动孢子囊和游动孢子，游动孢子具有双鞭毛。有的孢子囊能直接萌发形成芽管，发育成菌丝体。孢囊梗和孢子囊的形态是卵菌的重要分类依据。有性生殖产生的雌雄配子体高度分化，即藏卵器(♀)和雄器(♂)，经卵配生殖形成卵孢子。

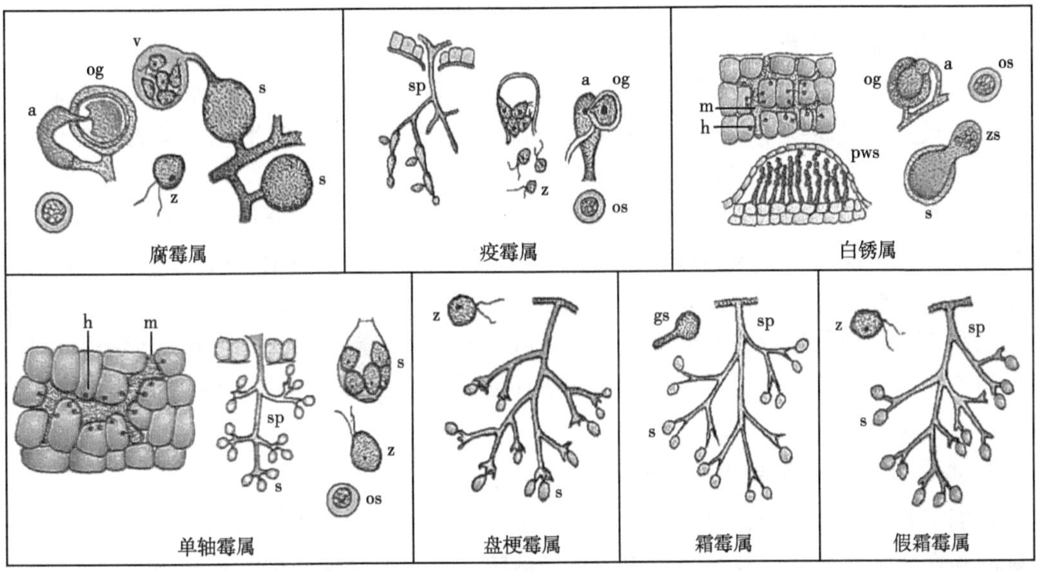

a. 雄器；gs. 萌发的孢子囊；h. 吸器；m. 菌丝体；og. 藏卵器；os. 卵孢子；pws. 孢囊堆；s. 孢子囊；sp. 孢囊梗；v. 泡囊；z. 游动孢子；zs. 游动孢子囊。

图 2-1　卵菌常见病原物的形态特征

(引自 Agrios，2005)

【实验目的】

1. 了解卵菌门菌物的主要形态特征，掌握与植物病害相关的重要属的基本形态特征与分类依据。

2. 掌握卵菌门病原物重要属所致植物病害的症状特点。

3. 学习临时显微玻片制作技术。

【材料和器具】

1. 实验材料

①腐霉科：黄瓜腐霉病(*Pythium aphanidermatum*)离体培养物；马铃薯晚疫病的标本(具霉层)以及病原物(*Phytophthora infestans*)的孢子囊、孢囊梗和卵孢子的显微玻片。

②霜霉科：葡萄霜霉病的标本(具霜霉层)以及病原物(*Plasmopara viticola*)的孢子囊和孢囊梗的显微玻片；大豆霜霉病的标本(具霜霉层)以及病原物(*Peronospora manschurica*)的

孢子囊和孢囊梗的显微玻片。

2. 实验器具

计算机、投影仪、显微镜、手持放大镜、解剖针、载玻片、盖玻片、镊子、剪刀、手术刀、吸水纸、刀片等。

【方法和步骤】

1. 卵菌门病原物的致病性分析及所致植物病害的症状观察

通过对供试植物病害标本的症状观察，掌握这几种不同卵菌代表属所致植物病害症状的特点，分析它们的寄生性和致病性的变化及关联。

2. 腐霉科病原物重要属的特征观察

①腐霉属(*Pythium*)：用尖头镊子取少许黄瓜腐烂病的培养物，以水为浮载剂制作成临时显微玻片。显微镜下观察其菌丝有无隔膜，孢子囊有无一定形状，有无分化的孢囊梗，注意泡囊的形状。边观察边绘制形态图。

②疫霉属(*Phytophthora*)：疫霉属的孢囊梗与菌丝有明显的分化，孢囊梗分枝基部呈节状膨大，并呈现连续生长的特点，使多个分枝排列在同一轴线上，故呈现假单轴分枝的特点。孢子囊柠檬形、球形或梨形，具有乳状突起。取马铃薯晚疫病病斑上的白色霉状物，以水为浮载剂制作成临时显微玻片。显微镜下观察其孢子囊和孢囊梗的形态。也可直接采用该属病原菌的永久显微玻片直接观察其孢子囊、孢囊梗和卵孢子的形态特征。边观察边绘制形态图。

3. 霜霉科病原物重要属的特征观察

分别取葡萄霜霉病和大豆霜霉病标本，用刀片轻轻刮下发病部位上的霜霉层，以水为浮载剂制作成临时显微玻片。显微镜下观察这两个属的孢囊梗和孢子囊形态特征。也可直接采用永久显微玻片直接观察其孢子囊和孢囊梗的形态特征。边观察边绘制形态图。

①单轴霉属(*Plasmopara*)：孢囊梗单轴分枝，分枝与主轴成直角或近直角，分枝顶端有2~3个短而钝的小梗，孢子囊着生在小梗上，孢子囊无色，卵形至椭圆形，顶端有突起。

②霜霉属(*Peronospora*)：孢囊梗主轴粗壮，顶部有多轮左右排列的二叉状分枝，分支末端顶端尖锐，孢子囊近卵形，成熟时易脱落，萌发时直接形成芽管，偶尔释放游动孢子。

【结果和讨论】

1. 结合腐霉科和霜霉科各重要属病原物的形态及所致植物病害症状特点，填写表2-2。

表2-2　卵菌重要病原物形态特征及所致病害特点

病菌属名	孢囊梗	孢子囊	寄生性	病害类型	危害部位
腐霉属					
疫霉属					
单轴霉属					
霜霉属					

2. 根据显微镜观察，分别绘制腐霉属、疫霉属、霜霉属和单轴霉属的孢囊梗和孢子囊的形态特征图，并做标注。

3. 根据孢子囊、孢囊梗和藏卵器特征以及寄生习性，编制霜霉目病原物重要属的检索表。

4. 卵菌与真菌有哪些区别？

EXPERIMENT 2　Recognition and Identification of Important Plant Pathogens of Oomycota

【Introduction】

Oomycota is one of the important plant pathogens. Peronosporales is the most important pathogen group of Oomycota, which can cause downy mildew, root rot, blight, damping off, seed mildew and white rust, ect.

The habits of Oomycota change from saprophytism to obligate parasitism and from hydrophilous to terrestrial. Correspondingly, the morphology of the reproductive body shows different characteristics, such as the sporangiophores showing from filamentous shape to distinct differentiation, the sporangia showing from non-shedding to shedding (Figure 2-1), the transmitting vectors changing from waterborne to airborne, the parasitic site changing from the soil surface to the junction between stem and root to the stems and leaves on the ground, and the symptoms changing from decay and necrosis to promoting symptom, and so on.

The vegetative bodies of Oomycota are developed septate mycelium. It produces zoosporangia and zoospores for asexual reproduction. Each zoospore has double flagella. Some sporangia can germinate directly to form germ tube and develop into mycelia. The morphology of sporangium and sporangiophore is the important classification basis of Oomycota. The male and female gametangia produced by sexual reproduction are highly differentiated, i. e. the oogonium(♀) and the antheidium(♂), which are via oogamy to produce oospores.

【Experimental Purpose】

1. Understand the main morphological characteristics of the fungi in Oomycota, and master the basic morphological characteristics and classification basis of the important pathogenic genera related to plant diseases in Oomycota.

2. Master the symptoms of plant diseases caused by the important pathogenic genera in Oomycota.

3. Learn the techniques of making temporary microslides.

【Materials and Apparatus】

1. Materials

①Pythiaceae: The *in vitro* culture of *Pythium aphanidermatum*, the pathogen of cucumber rot; the specimen of potato late blight covered by a mildew layer, and the microslides of sporangium,

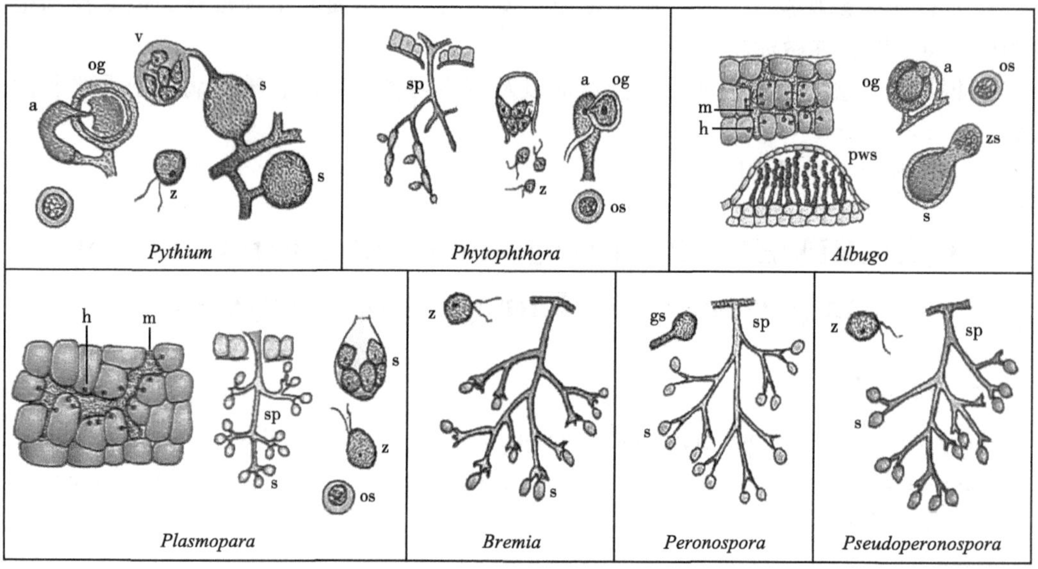

a. antheridium; gs. germinating sporangium; h. haustorium; m. mycelium; og. oogonium; os. oospore; pws. pustule with sporangia; s. sporangium; sp. sporangiophore; v. vesicle; z. zoospore; zs. zoosporangium.

Figure 2-1　Morphological Chracteristics of Common Pathogens in Oomycota

(Cited from Agrios, 2005)

sporangiophore and oospore of *Phytophthora infestans*.

②Peronosporaceae: The specimen of grape downy mildew covered by a mildew layer, and the microslides of sporangium and sporangiophore of *Plasmopara viticola*; the specimen of soybean downy mildew covered by a mildew layer, and the microslides of sporangium and sporangiophore of *Peronospora manschurica*.

2. Instruments and Appliances

Computers, projectors, microscopes, handheld magnifying glasses, dissecting needles, glass slides, coverslips, tweezers, scissors, scalpels, absorbent paper, blades, etc.

【Methods and Procedures】

1. Pathogenicity Analysis and Symptom Observation of Pathogens in Oomycota

By observing the symptoms of the provided plant disease specimens, master the characteristics of symptoms caused by the different representative genera of Oomycota, and analyze the change and correlation of their parasitism and pathogenicity.

2. Observation on Characteristics of Important Pathogen Genera of Pythiaceae

①*Pythium*: Pick up a little culture of *Pythium aphanidermatum* using pointed tweezers, and make temporary microslides using water as floating agent. And then observe its hyphae, septa, the shape of sporangium, the differentiation status of sporangiophore and the shape of vesicle under a microscope. Draw the morphological illustration as you observe.

②*Phytophthora*: The sporangiophore of *Phytophthora* are obviously differentiated from hyphae.

The base of sporangiophore expands into nodular structure. The sporangiophores show continuous growth to make many branches arrange on the same axis, so it shows pseudouniaxial branching. The sporangium of *Phytophthora* shows lemon-shaped globose or pear-shaped with the structure like papilla. Pick up the white mildew on the spot of potato late blight to make temporary microslides using water as floating agent, and observe the morphology of sporangium and sporangiophore under a microscope. The morphological characteristics of the sporangia, sporangiophores, and oospores can also be observed directly using the permanent microscopic slides of *Phytophthora infestans*. Draw the morphological illustration as you observe.

3. Observation on Characteristics of Important Pathogen Genera of Peronosporaceae

Taking the specimen of grape downy mildew and soybean downy mildew, scrape the downy mildew layer on the diseased site with a blade, make temporary microslides using with water as floating agent, and observe the morphological characteristics of sporangium and sporangiophore of the two genera under a microscope. The morphologic features of sporangium and sporangiophore can also be observed directly by permanent microscope slides. Draw the morphological illustration as you observe.

①*Plasmopara*: The sporangiophores are uniaxial branches. The branches form right angles or near right angles to the main shaft, and there are 2-3 short and blunt petioles on the tips of the branches. The sporangium is formed on the petioles, which is colorless, ovate to elliptic, and has a protuberance on its top.

②*Peronospora*: The main shaft of the sporangiophore is hairchested, and there are many bifurcate branches showing as bilateral arrangement on its tops. The tip of the terminal branch is sharp. The sporangium is nearly ovate and easy to fall off at maturity, which forms germ tubes directly when germinating, occasionally releasing zoospores.

【Results and Discussion】

1. Fill in Table 2-2 according to the morphological characteristics and symptoms of plant diseases caused by pathogens from each important genus of Pythiaceae and Peronosporaceae.

Table 2-2　Morphological Characteristics and Disease Symptoms of Important Pathogens in Oomycota

Pathogen	Sporangiophore	Sporangium	Parasitism	Disease types	Disease tissues
Pythium					
Phytophthora					
Plasmopara					
Peronospora					

2. Draw the morphological illustrations of the sporosporium and sporangiophore of *Pythium*, *Phytophthora*, *Peronospora* and *Plasmopara* and make notes according to the microscope observation.

3. List a retrieve table of the important pathogen genera of Peronosporales according to the characteristics of sporangia, sporangiophores, oogonium and the parasitic habits.

4. What are the differences between Oomycota and true fungi?

实验三　接合菌重要植物病原物的识别与鉴定

【概述】

接合菌门菌物的共同特征是有性生殖产生接合孢子，无性繁殖产生不能游动的孢囊孢子。接合菌在自然界分布广泛，大多数为腐生菌，腐生于土壤、植物残体、动物粪便和多种有机质上，少数是寄生菌，与植物病害相关的是毛霉目(Mucorales)(图 2-2)。

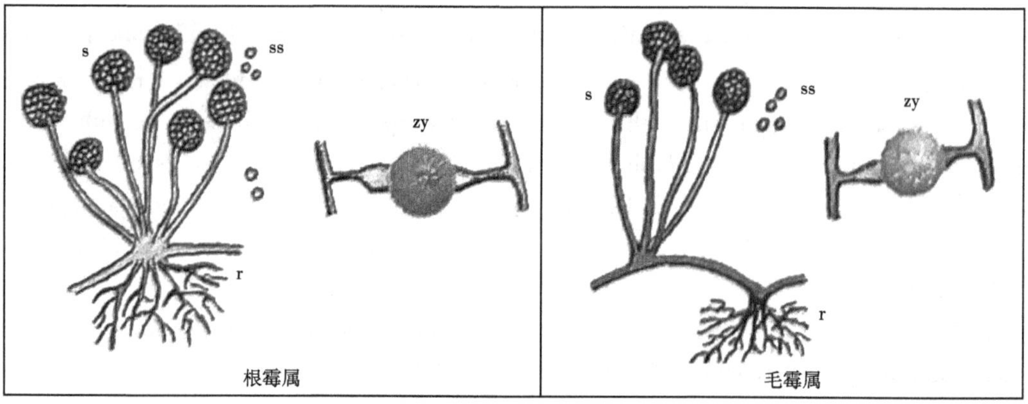

s. 孢子囊；ss. 孢囊孢子；zy. 接合孢子；r. 假根。

图 2-2　接合菌常见病原物的形态特征

(引自 Agrios，2005)

毛霉目是接合菌中比较重要的类群，其中绝大多数在土壤及有机质上附着，寄生于植物上的极少，最重要的种为匍枝根霉(*Rhizopus stolonifer*)，常导致薯类和水果发生软腐。此外，毛霉属(*Mucor*)和犁头霉属(*Absidia*)的某些种，也能引起储藏谷物的腐烂。

【实验目的】

1. 了解接合菌门菌物的主要形态特征，掌握与植物病害相关的重要属的基本形态特征和分类依据。
2. 了解接合菌重要属病原物所致植物病害症状的特点。
3. 熟悉临时显微玻片的制作技术。

【材料和器具】

1. 实验材料

匍枝根霉的离体培养物；甘薯软腐病、果实腐烂病和花瓣腐烂病的标本(具霉层)；接合孢子和孢囊孢子的永久显微玻片。

2. 实验器具

计算机、投影仪、显微镜、手持放大镜、解剖针、载玻片、盖玻片、镊子、剪刀、手术刀、吸水纸、刀片等。

【方法和步骤】

1. 用显微镜直接观察接合孢子和孢囊孢子的永久显微玻片。边观察边绘制形态图。

2. 用尖头镊子挑取少许匍枝根霉培养物，以水为浮载剂制作临时显微玻片，在显微镜下观察其孢子囊、孢囊梗、假根及匍匐丝的形态。边观察边绘制形态图。

3. 观察供试植物病害标本，总结根霉属病原物所致植物病害的症状特点。

【结果和讨论】

1. 绘制接合孢子的形态图。

2. 根据显微镜观察，绘制匍枝根霉的孢子囊、孢囊梗、假根及匍匐丝的形态特征图，并做标注。

3. 简要描述接合菌有性生殖和无性繁殖的过程。

EXPERIMENT 3　Recognition and Identification of Important Plant Pathogens of Zygomycota

【Introduction】

The common characteristic of Zygomycota is that it produces zygospore for sexual reproduction and produces immotile sporangiospores for asexual reproduction. Zygomycota is widely distributed in nature. Most of Zygomycota are saprophytes, which are usually found in soil, plant residues, animal feces and a variety of organic matter. A few of Zygomycota are parasitic, and the most related order with plant diseases is Mucorales(Figure 2-2).

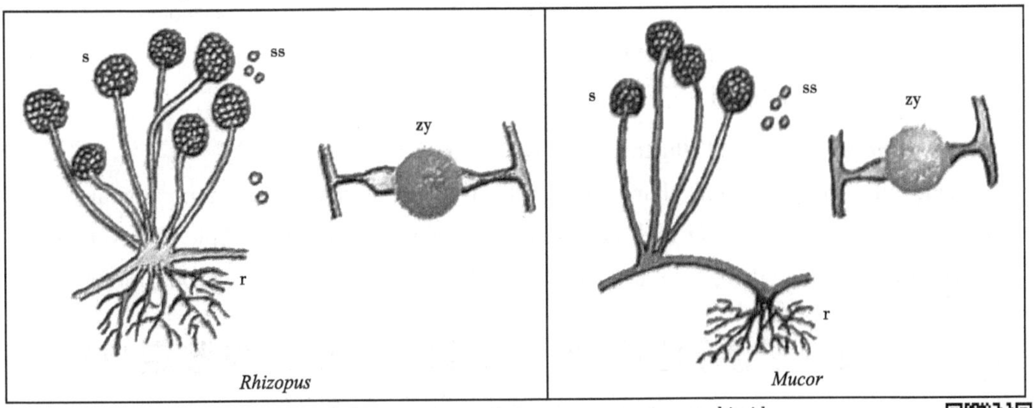

s. sporangium; ss. sporangiospore; zy. zygospore; r. rhizoid.

Figure 2-2　Morphological Characteristics of Common Pathogens in Zygomycota

(Cited from Agrios, 2005)

Mucorales is an important group of Zygomycota, most of which are attached to soil and organic matter, while a few are parasitic in plants. The most important species *Rhizopus stolonifer*, which usually causes soft rot of tube crops and fruits. In addition, some species of the genera *Mucor* and *Absidia* can also result in rot of stored cereal.

【Experimental Purpose】

1. Understand the main morphological characteristics of Zygomycota, and master the basic morphological characteristics and classification basis of important genera related with plant diseases.

2. Understand the symptom characteristics of plant disease caused by the important pathogen genera of Zygomycota.

3. Be familiar with the technique of making temporary microslides.

【Materials and Apparatus】

1. Experimental Materials

The culture of *Rhizopus stolonifera in vitro*, the specimens of sweet potato soft rot, fruit rot and petal rot (with moldy layer), and the permanent microscopic slides of zygospores and sporangiospores.

2. Instruments and Appliances

Computers, projectors, microscopes, handheld magnifying glasses, dissecting needles, slides, cover glasses, tweezers, scissors, scalpels, absorbent paper, blades, etc.

【Methods and Procedures】

1. Observe the permanent microscopic slides of zygospore and sporangiospore directly under a microscope. Draw the morphological illustrations as you observe.

2. Pick up a little culture of *Rhizopus stolonifer* with pointed tweezers and make temporary microslides using water as floating agent, and observe the morphology of the sporangia, sporangiosphore, rhizoid and stolon under a microscope. Draw the morphological illustrations as you observe.

3. Observe the provided specimens of plant diseases, and summarize the symptom characteristics of plant diseases caused by the pathogens of *Rhizopus*.

【Results and Discussion】

1. Draw the morphological illustration of zygospore.

2. Draw the morphological characteristic illustrations of the sporangium, sporangiophore, rhizoid and stolon of *R. stolonifer* and make notes according to the microscopic observation.

3. Describe the sexual reproduction and asexual reproduction of Zygomycota briefly.

实验四　子囊菌重要植物病原物的识别与鉴定(I)
——外囊菌目重要属形态观察

【概述】

子囊菌门(Ascomycota)是菌物中最大的类群，其共同特征是有性生殖形成子囊和子囊孢子，但它们在形态、生活史和生活习性上有很大的差别。Ainsworth(1973)的分类系统将子囊菌门分为6个纲：半子囊菌纲、不整囊菌纲、核菌纲、腔菌纲、盘菌纲和虫囊菌纲，其中虫囊菌纲主要危害昆虫，其余5个纲的真菌均与植物病害相关。子囊菌分纲的主要依据是有性阶段的特征，包括是否形成子囊果、子囊果的类型和子囊的特征等。

半子囊菌纲(Hemiascomycetes)不形成子囊果，子囊外面没有包被，营养体是单细胞或是很不发达的菌丝体；有性生殖比较简单，不形成特殊分化的配子囊，子囊也不是由产囊丝形成的，而是由菌丝细胞直接形成。与植物病害关系密切的外囊菌目(Taphrinales)只有1科1属，即外囊菌科外囊菌属(*Taphrina*)，都是蕨类或高等植物的寄生菌，可引起叶片、枝梢和果实的畸形(图2-3)。

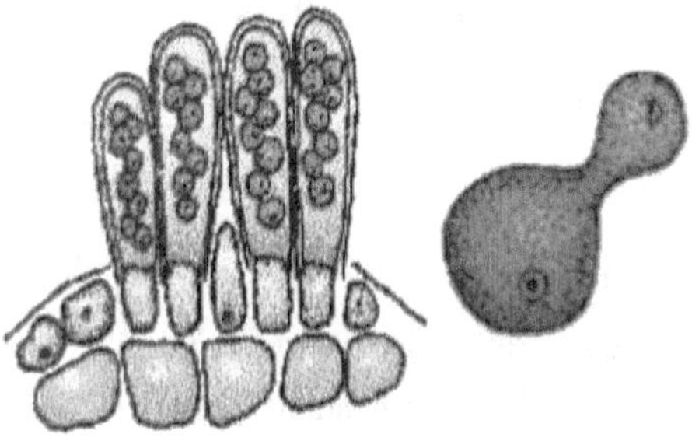

外囊菌属
（引自Agrios，2005）

桃缩叶病
（引自Moscatello et al.，2017）

图2-3　外囊菌属的形态特征及其引起的植物病害症状

【实验目的】

1. 掌握外囊菌属的重要形态特征。
2. 观察外囊菌属所致植物病害的症状特点，并掌握其重要的鉴定依据。
3. 学习徒手切片技术。

【材料和器具】

1. 实验材料

①植物病害标本：桃缩叶病、杏缩叶病和樱桃丛枝病(具白色粉蜡层)。

②永久显微玻片：桃缩叶病(*Taphrina deformans*)的子囊和子囊孢子。

2. 实验器具

计算机、投影仪、显微镜、手持放大镜、解剖针、载玻片、盖玻片、镊子、剪刀、手术刀、吸水纸、刀片、通草(使用前浸泡在70%乙醇中)等。

【方法和步骤】

1. 用显微镜直接观察桃缩叶病示范切片的子囊和子囊孢子形态特征，边观察边绘制形态图。

2. 观察供试植物病害标本的症状，记录外囊菌所致植物病害症状特点。

3. 从供试植物病害标本中任选一种植物病害，取具有明显病征(具白色粉蜡层)的小块发病叶片组织，以通草作为固定媒介，制作徒手切片，然后在显微镜下观察病菌的子囊和子囊孢子，边观察边绘制形态图。

【结果和讨论】

1. 绘制桃缩叶病病原菌的子囊和子囊孢子形态图，并做标注。
2. 总结外囊菌属的重要形态特征及其所致植物病害症状特点。

EXPERIMENT 4 Recognition and Identification of Important Plant Pathogens of Ascomycota(Ⅰ)

——Morphological Observation of Important Genera in Taphrinales

【Introduction】

Ascomycota is the largest group of fungi, which is characterized by the formation of ascus and ascospore for sexual reproduction. However, the morphology, life history and living habits show dramatic differences among the fungi of Ascomycota. According to the classification system of Ainsworth (1973), Ascomycota was divided into 6 classes: Hemiascomycetes, Plectomycetes, Pyrenomycetes, Loculoascomycetes, Discomycetes and Laboulbeniomycetes, of which Laboulbeniomycetes is mainly harmful to insects, while all the other five classes are related to plant diseases. The classification of Ascomycota on the class level is mainly based on its characteristics of sexual stage, that is, whether to form ascocarp, the types of ascocarps, and the characteristics of asci, etc.

The fungi of Hemiascomycetes can't form ascocarp, and there are no any envelops outside the asci. The vegetative bodies of Hemiascomycetes are unicellular or poorly developed mycelium. The sexual reproduction of Hemiascomycetes is relatively simple and can't form special differentiated gametangia. The asci are directly formed by hyphal cell instead of ascogenous hypha. Taphrinales is the order that is the most closely related to plant diseases, only including one family one genus, i. e. *Taphrina*. *Taphrina* mainly parasitize in ferns or higher plants to cause the deformities in leaves, shoots and fruits(Figure 2-3).

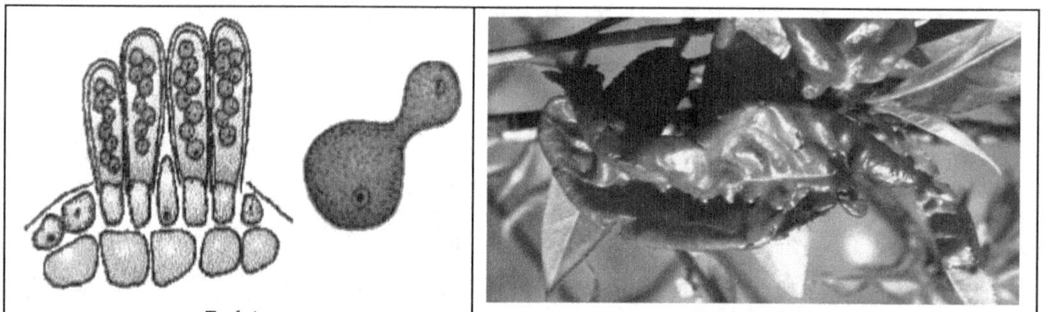

Taphrina (Cited from Agrios, 2005)

Peach fern leaf (Cited from Moscatello et al., 2017)

Figure 2-3 Morphological Characteristics of *Taphrina* and the Symptom of Plant Disease Caused by It

【Experimental Purpose】

1. Master the important morphological characteristics of *Taphrina*.
2. Observe the symptom characteristics of plant diseases caused by *Taphrina*, and grasp its

important identification basis.

3. Master the technique of hand-making section.

【Materials and Apparatus】

1. Materials

①Plant disease specimens: Fern leaf of peach, fern leaf of apricot, and withes broom of cherry (with white powdery wax layer).

②Permanent microscopic slides: The asci and ascospores of the pathogen of peach fern leaf (*Taphrina deformans*).

2. Instruments and Appliances

Computers, projectors, microscopes, handheld magnifying glasses, dissecting needles, slides, cover glasses, scissors, scalpels, blotting papers, blades, ricepaperplant piths(soaked in 70% alcohol before use), etc.

【Methods and Procedures】

1. Observe the asci and ascospores of *T. deformans* on the model slides directly using microscope. Draw the morphological illustrations as you observe.

2. Observethe symptoms of plant disease specimens, and record the symptom characteristics of the plant diseases caused by *Taphrina*.

3. Select any plant disease among all the provided specimens, take one small patch of diseased leaf with obvious symptom(covered by a white wax powdery layer) to make sections using pith as the fixing medium, and then observe the asci and ascospores of the pathogen under the microscope. Draw the morphological illustrations as you observe.

【Results and Discussion】

1. Draw the morphological illustrations of the asci and ascocpores of *T. deformans*, and make notes.

2. Summarize the important morphological characteristics of *Taphrina* and the symptom characteristics of plant diseases caused by it.

实验五　子囊菌重要植物病原物的识别与鉴定(Ⅱ)
——白粉菌目重要属形态观察

【概述】

核菌纲(Pyrenomycetes)是子囊菌中最大的纲，其重要的形态特征是子囊果为有孔口的子囊壳，子囊着生在子囊壳内；或形成无孔口的闭囊壳，子囊有规律地排列在闭囊壳内。Ainsworth(1973)的分类系统将核菌纲分为4个目：白粉菌目、小煤炱目、冠囊菌目和球壳目，其中与植物病害最相关的是白粉菌目和球壳目。

白粉菌目(Erysiphales)真菌一般称为白粉菌，都是高等植物的专性寄生菌，可引起植物的白粉病。白粉病的典型症状为发病前期在发病组织表面产生白色的粉层，即白粉菌的菌丝体、分生孢子和分生孢子梗，后期在白色粉层上会产生黑色的肉眼可见的小黑粒，即闭囊壳。闭囊壳是白粉菌进行有性生殖形成的子囊果，成熟的闭囊壳球形或近球形，四周或顶端有各种形状的附属丝。闭囊壳内有 1 个或多个子囊，子囊有卵形、椭圆形或圆筒形，子囊中有 2~8 个椭圆形的子囊孢子。附属丝的形态及闭囊壳内子囊的数量是白粉菌分类鉴定的重要依据。

【实验目的】

1. 了解白粉菌的分类地位，掌握白粉菌的重要形态特征。
2. 掌握植物白粉病的症状特点，以及引起白粉病的重要病原物属的主要鉴定特征。
3. 熟悉临时显微玻片制作技术。

【材料和器具】

1. 实验材料

植物病害标本：芍药白粉病(*Erysiphe paeoniae*)、凤仙花白粉病(*Sphaerotheca balsaminae*)、刺槐白粉病(*Microsphaera subtrichotoma*)、桃树白粉病(*Podosphaera tridatyla*)、黄栌白粉病(*Uncinula verniciferae*)和桑白粉病(*Phyllactinia moricola*)。所提供的植物病害标本应为病征明显的发病植物部位，即能观察到明显的白粉层和黑色的小点粒。

2. 实验器具

计算机、投影仪、显微镜、手持放大镜、解剖针、载玻片、盖玻片、镊子、剪刀、刀片、手术刀、吸水纸等。

【方法和步骤】

1. 仔细观察所提供的植物白粉病标本的病状和病征，总结植物白粉病的症状特点。
2. 选取供试植物标本具有明显白粉病病征的发病部位，用刀片在发病部位轻轻刮取黑色的小点粒(即白粉病病原菌的闭囊壳)，转移至载玻片上，以水为浮载剂制作成临时显微玻片，在显微镜下观察闭囊壳外周的附属丝形态特点，然后用镊子轻轻压盖玻片，使闭囊壳破裂，释放出子囊，仔细观察闭囊壳内的子囊数量和形状(图 2-4)。
3. 参照检索表 2-3，从附属丝形态差异和子囊数量区分白粉病病原物的重要属，并结合显微镜观察结果绘制其形态图。

表 2-3　白粉菌目常见重要病原菌属检索表

1. 闭囊壳内单子囊
　2. 附属丝菌丝状 …… 单丝壳属
　2. 附属丝刚直，顶端二叉状分枝 …… 叉丝单囊壳属
1. 闭囊壳内多子囊
　2. 附属丝菌丝状 …… 白粉菌属
　2. 附属丝非菌丝状 …… 3
　　3. 附属丝基部球状膨大，顶端针状 …… 球针壳属
　　3. 附属丝非球针状 …… 4
　　　4. 附属丝刚直，顶端二叉状分枝 …… 叉丝壳属
　　　4. 附属丝刚直，顶端钩状 …… 钩丝壳属

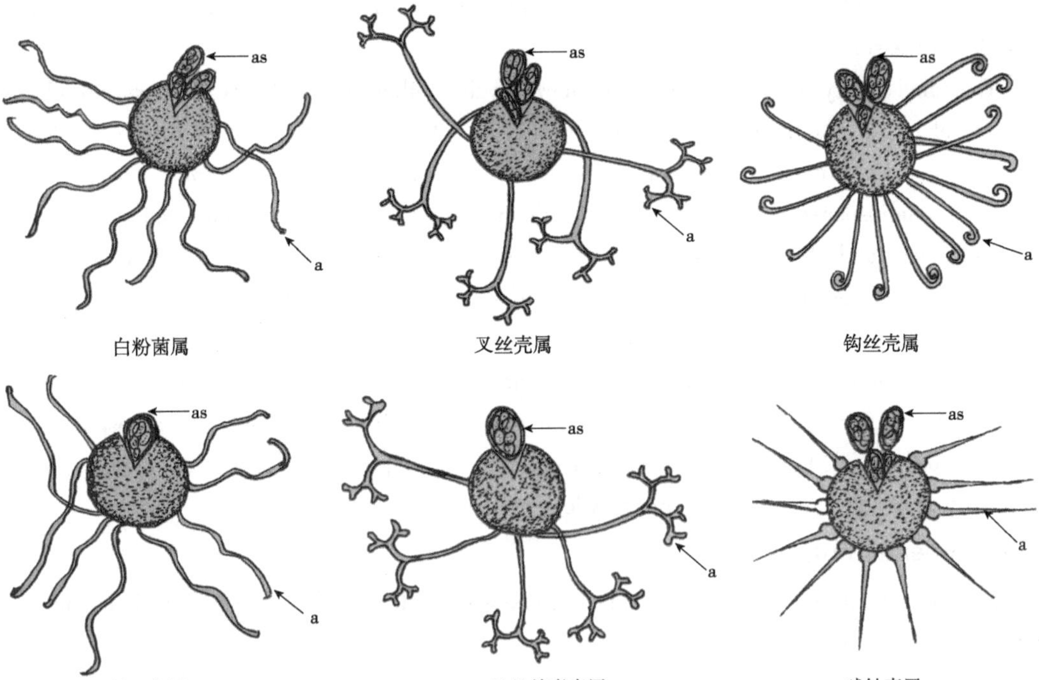

单丝壳属　　叉丝单囊壳属　　球针壳属

a. 附属丝；as. 子囊。

图 2-4　6 种重要的白粉病病原菌的闭囊壳形态图

【结果和讨论】

1. 绘制白粉菌属、单丝壳属、叉丝单囊壳属、叉丝壳属、球针壳属和钩丝壳属的闭囊壳的形态图，并做标注。
2. 谈谈白粉菌的主要分类依据。
3. 谈谈白粉菌目和小煤炱目真菌在寄生性和形态特征上的异同。
4. 简述植物白粉病的主要诊断依据。

EXPERIMENT 5　Recognition and Identification of Important Plant Pathogens of Ascomycota (Ⅱ)

——Morphological Observation of Important Genera in Erysiphales

【Introduction】

Pyrenomycetes is the largest class of Ascomycota. The most important morphological feature of Pyrenomycetes is that its ascocarp is perithecium with asci inside it, or, it forms cleistothecium without orifice with asci arranged regularly inside it. According to the taxonomy system of Ainsworth (1973), Pyrenomycetes was divided into four orders, i. e. Erysiphales, Melioales,

Coronophorales and Sphaeriales, of which Erysiphales and Sphaeriales are the most related orders with plant diseases.

The fungi of Erysiphales, commonly known as the pathogen of powdery mildew, are obligate parasites of higher plants and cause powdery mildew. The typical symptom of powdery mildew is that there is a white powdery layer on the surface of diseased tissues at the early stage, i. e. the mycelia, conidiospores and conidiophores of the pathogen, while at the later stage some black and macroscopical little particles are produced on the surface of the white powdery layer, i. e, cleistothecium. Cleistothecium is the asocarp that is formed when the fungi in Erysiphales conducts sexual reproduction. The mature cleistothecium is spherical or subspherical, and with various appendages around it or on its top. There are one or more asci in each cleistothecium, and the asci are usually ovoid, elliptical or cylindrical in shape with two to eight oval ascospores in it. The morphological characteristics of cleistothecium are the important basis for the classification and identification of the pathogens causing powdery mildew.

【Experimental Purpose】

1. Understand the taxonomic status of powdery mildew fungi, and grasp their important morphological characteristics.

2. Master the symptom characteristics of plant powdery mildew and the main identification characteristics of the important pathogenic genera of powdery mildew.

3. Be familiar with the technique of making temporary microslides.

【Materials and Apparatus】

1. Materials

Plant disease specimens: Powdery mildew of poeny(*Erysiphe paeoniae*), powdery mildew of balsamine(*Sphaerotheca balsaminae*), powdery mildew of locust(*Microsphaera subtrichotoma*), powdery mildew of peach (*Podosphaera tridatyla*), powdery mildew of smoke tree (*Uncinula verniciferae*) and powdery mildew of mulberry (*Phyllactinia moricola*). All the provided plant disease samples should be the diseased tissues with obvious sign, i. e. the distinct white powder layers and black little particles can be observed.

2. Instruments and Appliances

Computers, projectors, microscopes, handheld magnifying glasses, dissecting needles, slides, cover glasses, tweezers, scissors, blades, scalpels, blotting papers, etc.

【Methods and Procedures】

1. Observe the morbidity and sign of the specimens of plant powdery mildew carefully, and summarize the symptom characteristics of plant powdery mildew.

2. Select the diseased sites of the provided speciemens of plant powdery mildew with obvious sign, and then use a blade to scrape the black little particles gently from the diseased sites, i. e. the cleistothecia of powdery mildew fungi, which are then transferred to the slide and made as temporary microscope slides using water as floating agent. Employ a microscope to observe the

morphological characteristics of appendages around the cleistothecia, then use a tweezers to press the slide gently to crack the cleistothecium to release asci, and observe the number and shape of ascus in cleistothecium(Figure 2-4).

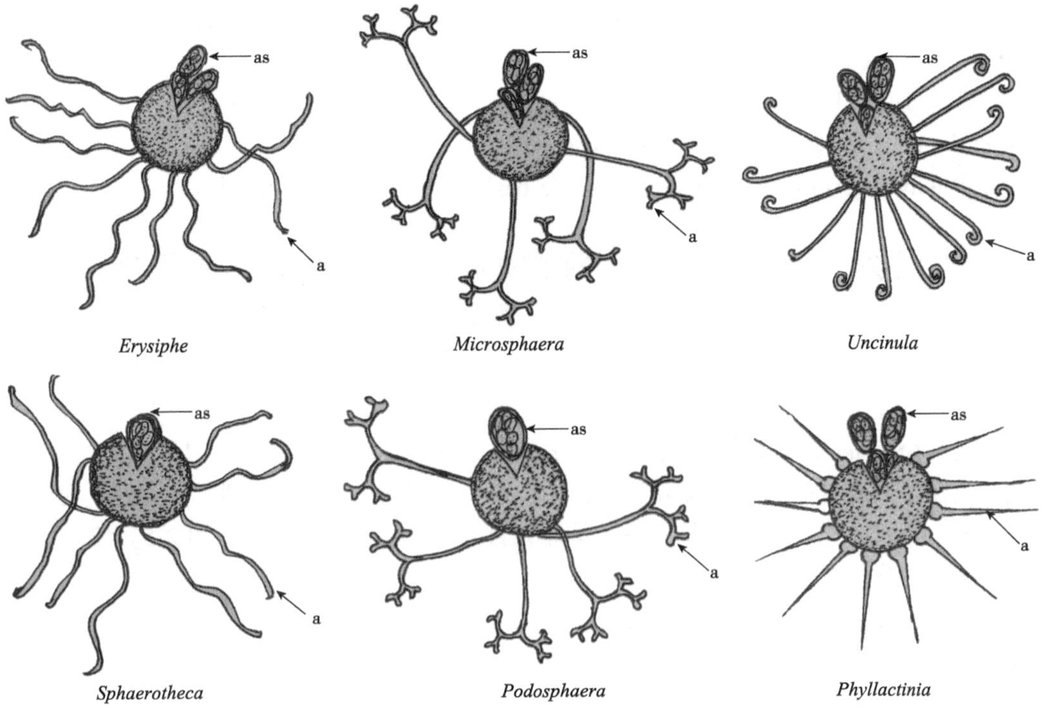

a. appendage; as. ascus.

Figure 2-4　Morphological Illustrations of Cleistothecia of Six Types of Important Pathogens Causing Powdery Mildew

3. Refer to the retrieve Table 2-3, distinguish the important pathogenic genera of powdery mildew based on the morphological differences of appendages and the number of asci, and draw the morphological illustrations of them combining with observation results using a microscope.

Table 2-3　Retrieve Table of the Common and Important Pathogenic Genera in Erysiphales

1. A single ascus in the cleistothecium
    2. Appendages are filamentous ······ *Sphaerotheca*
    2. Appendages are upright with a top like bifurcate branching ······ *Podosphaera*
1. Multiple ascus in the cleistothecium
    2. Appendages are filamentous ······ *Erysiphe*
    2. Appendages are not filamentous ······ 3
        3. The base of appendage is bulbous with a needle-like top ······ *Phyllactinia*
        3. Appendages are not bulbous needle-like in shape ······ 4
            4. Appendages are upright with a top like bifurcate branching ······ *Microsphaera*
            4. Appendages are upright with a top like a hook ······ *Uncinula*

【Results and Discussion】

1. Draw the morphological illustrations of the cleistothecia of *Erysiphe*, *Sphaerotheca*, *Podosphaera*, *Microsphaera*, *Phyllactinia and Uncinula*, and make notes.

2. Talk about the main basis for the classification of the pathogens causing powdery mildew.

3. Talk about the similarities and differences of the parasitism and morphological characteristics between the fungi in Erysiphales and Melioales.

4. Briefly demonstrate the main diagnostic basis for plant powdery mildew.

实验六　子囊菌重要植物病原物的识别与鉴定(Ⅲ)
——球壳目重要属形态观察

【概述】

球壳目(Sphaeriales)是核菌纲(Pyrenomycetes)中最大的目，核菌纲中子囊果类型为子囊壳的真菌都归在球壳目中。球壳目真菌的形态变化很大，子囊壳的形状、颜色、孔口和着生方式，以及不孕丝状体的差异往往是球壳目真菌的重要区分依据(图 2-5)。球壳目真菌一般都有很发达的分生孢子阶段。

球壳目真菌大多为腐生，其中也有不少是寄生的，能引起重要的植物病害。

【实验目的】

1. 了解球壳目重要病原物引起的植物病害症状特点。

2. 掌握球壳目重要病原物属的形态鉴定特征。

3. 熟悉徒手切片制作技术。

【材料和器具】

1. 实验材料

①植物病害标本：苹果树腐烂病(*Valsa mali*)、花椒干腐病(*Gibberella pilicaris*)、苹果炭疽病(*Glomerella ciningulaia*)和竹叶黑痣病(*Phyllachora orbicula*)。

②永久显微玻片：赤霉属的子囊壳、子囊和子囊孢子。

2. 实验器具

计算机、投影仪、显微镜、手持放大镜、解剖针、载玻片、盖玻片、通草、刀片、镊子、剪刀、手术刀、吸水纸等。

【方法和步骤】

1. 观察供试植物病害标本的症状，记录球壳目子囊菌引起植物病害的症状特点。

2. 观察球壳目重要病原菌属的显微形态特征，并加以区分。

①黑腐皮壳属(*Valsa*)：黑腐皮壳属真菌大多为弱寄生菌，可引起多种树木的枝干腐烂病。病菌在扩展过程中，菌丝体可在寄主树皮内集结成青色颗粒状的子座。有性态形成子囊壳，着生在子座基部深处，长颈的孔口露出树皮外。子囊散生于子囊壳内，顶壁较

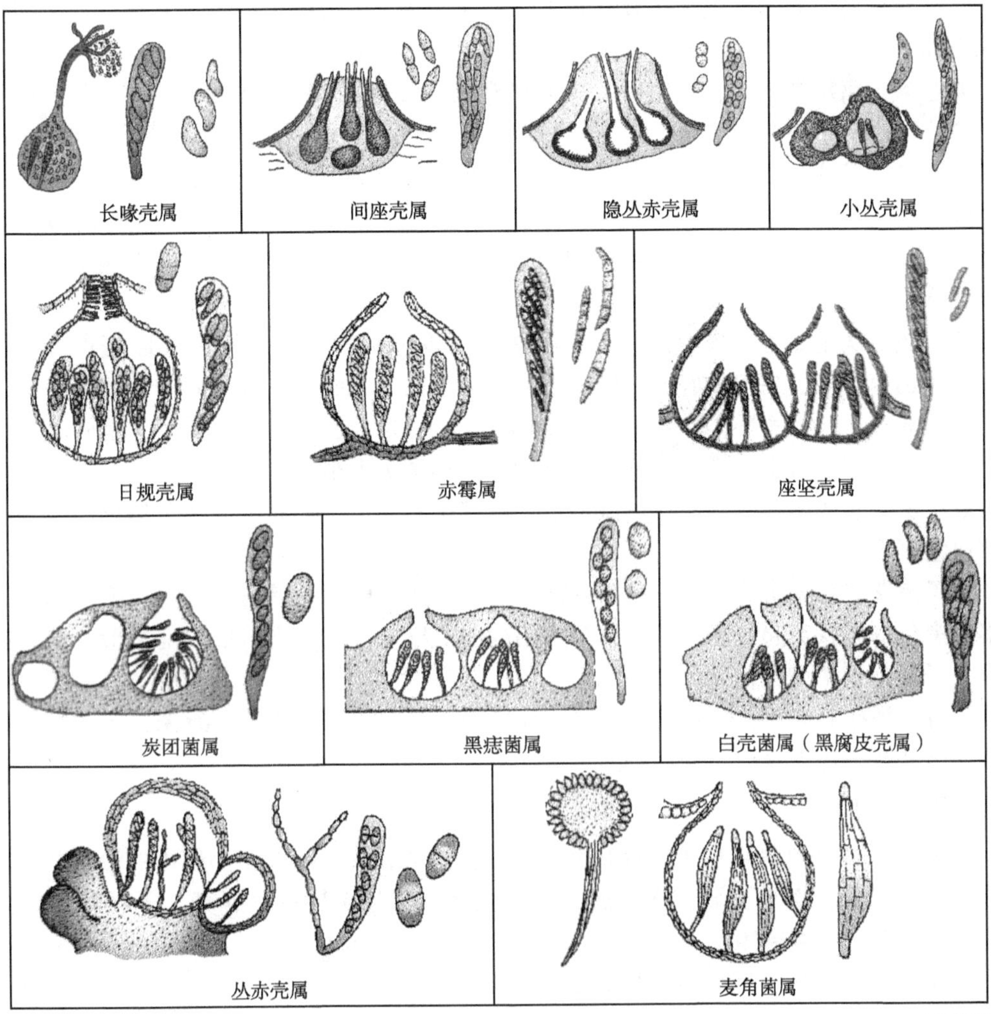

图 2-5　球壳目重要病原物的形态特征

（引自 Agrios，2005）

厚，棍棒状，内含 8 个子囊孢子，子囊孢子单孢，无色，香蕉形。有性态较少见，在病害的传播中的作用不明显，主要依靠无性繁殖产生大量的分生孢子进行传播。其无性态为壳囊孢属（*Cytospora*），分生孢子器产生在子座内，形状不规则，有长颈，孔口露出子座外，可以有几个相通的腔室，一个共同的孔口。分生孢子器内壁产生许多无色分生孢子梗，顶端着生分生孢子，分生孢子单细胞，无色，香蕉形，但较子囊孢子小。分生孢子与胶质混在一起，雨后或湿度很高时，吸水膨胀后自孔口挤出，呈橘黄色或橘红色的丝状物，干燥后硬化形成分生孢子角。

取病征明显的苹果腐烂病的树皮，切取一小块带有明显颗粒状子座的树皮病组织，以通草为固定媒介，制作徒手切片，在显微镜下观察黑腐皮壳属的子囊壳、子囊和子囊孢子的形态特征。边观察边绘制形态图。

②赤霉属（*Gibberella*）：赤霉属真菌常寄生在植物的茎、花器或种子上，引起赤霉病或

干腐病。赤霉属子囊壳表生，多聚集着生于垫状子座上，壳壁呈蓝色或紫色；子囊孢子具2~3个隔膜，无色或淡色，梭形，无性型为镰刀菌属(*Fusarium*)。

取花椒干腐病老病斑，先用手持放大镜观察病皮表面着生的蓝紫色颗粒——子囊壳，再用刀片取小块病皮制作徒手切片(或直接刮下子囊壳)或取该属的永久玻片，在显微镜下观察子囊壳、子囊和子囊孢子的形态特征。边观察边绘制形态图。

③小丛壳属(*Glomerella*)：小丛壳属真菌能引起多种植物的炭疽病，无性阶段为炭疽菌属(*Colletotrichum*)。子囊壳丛生于子座内，孔口突起，成熟时外露，没有侧丝，子囊孢子单细胞、无色。

苹果炭疽病主要为害果实，果实发病初期呈现针头大小的淡褐色圆形斑点，边缘清晰，病斑迅速扩大后果肉软腐，病斑下陷，病斑表面呈现颜色深浅相间的轮纹。当病斑扩大为1~2 cm时，自病斑中心生出突起的小粒点，初为褐色，后变为黑色，呈同心轮纹状排列，此即病菌的分生孢子盘。其有性阶段在自然情况下很少发现。

取小丛壳属的永久玻片，在显微镜下观察其子囊壳、子囊和子囊孢子的形态特征。边观察边绘制形态图。

④黑痣菌属(*Phyllachora*)：黑痣菌属真菌引起多种植物的黑痣病。发病组织部位，可观察到黑色的垫状突起，呈黑痣状，为病菌的假子座。子囊壳近球形或扁平形，群生在子座内，子囊孢子单细胞，无色。

取竹叶黑痣病病叶，制作徒手切片，显微镜下观察该属病菌的子囊壳、子囊和子囊孢子的形态特征。边观察边绘制形态图。

【结果和讨论】

1. 总结供试植物病害标本的症状特点，掌握球壳目真菌引起植物病害的诊断要点。

2. 分别绘制黑腐皮壳属、赤霉属、小丛壳属和黑痣菌属的子囊壳、子囊和子囊孢子的显微形态图，并做标注。

3. 编制一检索表，区分球壳目的重要植物病原物。

EXPERIMENT 6　Recognition and Identification of Important Plant Pathogens of Ascomycota (Ⅲ)

——Morphological Observation of Important Genera in Sphaeriales

【Introduction】

Sphaeriales is the largest order of Pyrenomycetes. All the fungi with the ascocarp as perithecium are categorized into Sphaeriales. The morphology of the fungi in Sphaeriales varies greatly, and the differences in the shape, color, orifice and growth mode of the perithecium, as well as the differences in the sterile filaments, are often the important basis for distinguishing them (Figure 2-5). The fungi in Sphaeriales usually have an extremely developed conidial stage.

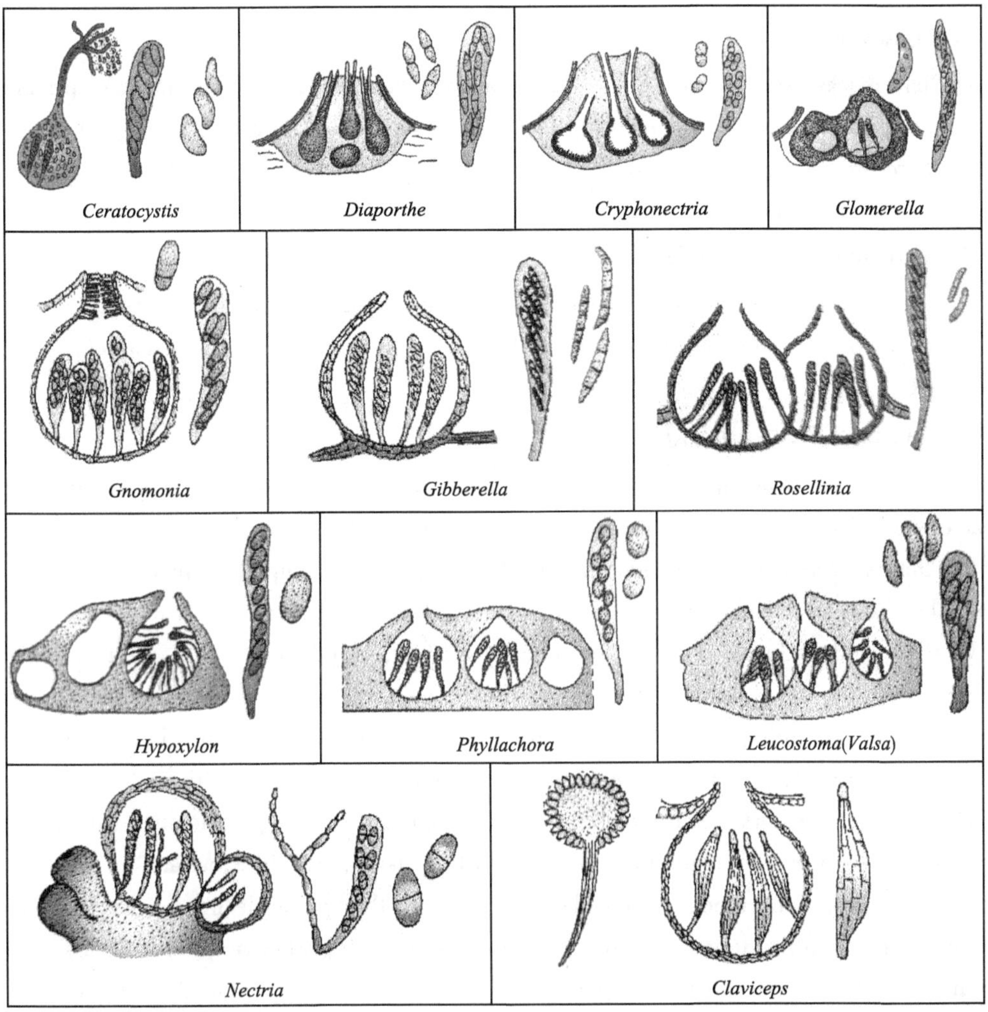

Figure 2-5　Morphological Characteristics of Important Pathogens in Sphaeriales

(Cited from Agrios, 2005)

Most of fungi in Sphaeriales are saprophytic, and many of them are parasitic, which can cause important plant diseases.

【Experimental Purpose】

1. Understand the symptom characteristics of plant diseases caused by the important pathogens of Sphaeriales.

2. Master the morphological identification features of the important pathogenic genera of Sphaeriales.

3. Be familiar with the technique of hand-making section.

【Materials and Apparatus】

1. Materials

①Plant disease specimens: Apple *Valsa* canker(*Valsa mali*), stem canker of prichly ash (*Gibberella pilicaris*), apple anthracnose(*Glomerella ciningulaia*) and black scurf of bamboo leaf (*Phyllachora orbicula*).

②Permanent microscopic slides: The perithecium, ascus and ascospore of *Gibberella.*

2. Instruments and Appliances

Computers, projectors, microscopes, handheld magnifying glasses, dissecting needles, slides, coverslips, ricepaperplant piths, blades, tweezers, scissors, scalpels, blotting paper, etc.

【Methods and Procedures】

1. Observe the symptoms of all provided plant diseases species, and record the symptom characteristics of the plant diseases caused by the fungi of Sphaeriales.

2. Observe the microscopic morphological features of the important pathogenic genera in Sphaeriales, and distinguish them.

① *Valsa*: Most fungi of *Valsa* are weak parasites, which can cause stem rot of many kinds of trees. In the expansion process of the pathogen, the cyan granular stroma can be formed by the assembling of mycelia in the bark of host plant. Perithecium is formed for teleomorph, which locates in the deep site of stroma base with the orifice of the long neck exposing outside the bark. The clavate asci are dispersed in the perithecium with thick top wall, in which there are eight unicellular colorless ascospores like banana in shape. The teleomorph is rarely observed, which plays little role in the spread of the disease. The spread of disease is mainly dependent on the abundant conidiospores produced by asexual reproduction. The anamorph is *Cytospora*, and the pycnidium is produced in the stroma with an irregular shape and a long neck with an orifice exposing outside the stroma, which can have some interlinked locules with the same orifice. Many colorless conidiophores grow on the inner cell of pycnidium. The conidiospore is produced on the top of conidiophore, which is unicellular, colorless, banana-like in shape and smaller than ascospore. The conidiospores are mixed with gelatin, which can swell after the rain or under a high moisture to squeeze out from the orifice to form yellow or orange filamentous structures and then become conidial angle when dried and harden.

Take the bark of apple *Valsa* canker with obvious sign, cut a small piece of bark tissue with obvious granular stroma, make hardworded sections using the pith as fixing medium, and then observe the morphological characteristics of perithecium, asci and ascospore of *Valsa* under a microscope. Draw the morphological illustrations as you observe.

②*Gibberella*: The fungi of *Gibberella* often parasitize in the stems, flower organs or seeds of plants to cause gibberellic disease or stem rot. The perithecia of *Gibberella* are produced in cluster on the surface of the cushion-like stroma. The wall of perithecium is blue or purple. The ascospore, with two to three septa, is colorless or pale, and fusiform. The anamorph of *Gibberella* is *Fusarium*.

Take the old spots of prickly ash stem canker, observe the blue-purple granules, i. e. perithecium, on the surface of the diseased bark using a handheld magnifying glass, and then take a small piece of diseased bark with a blade to make hardworked sections or directly scrape off the perithecium, or take the permanent glass slides of *Gibberella*, to observe the morphological characteristics of perithecium, asci and ascospore of *Gibberella* under a microscope. Draw the morphological illustrations as you observe.

③*Glomerella*: The fungi of *Glomerella* can cause anthracnose of a variety of plants with the anamorph as *Colletotrichum*. The perithecia are produced in cluster in the stroma with protuberant orifice, which are exserted at maturity. There is no paraphysis in the perithecium. The ascospore is unicellular and colorless.

Apple anthracnose mainly harms the fruit. At the initial stage of pathogenesis, the hazel round spots on the fruits are observed, which are pinhead-like in size with clear edges. As the rapid expansion of the spot, the fruit become soft rot, the spots are sunken and some alternating annuli of dark and light shade stripes are shown on the surface of spots. When the spots enlarge to 2-3 cm, some small protrusions appeared on the center of spots, which are initially brown and then black and arranged in concentric rings, i. e. the acervuli of the pathogen. The perfect stage of the pathogen is rarely found at natural conditions.

Take the permanent glass slides of *Glomerella*, and then observe the morphological characteristics of perithecium, asci and ascospore of *Gibberella* under a microscope. Draw the morphological illustrations as you observe.

④*Phyllachora*: The fungi of *Phyllachora* can cause black scurf of a variety of plants. On the diseased tissues, some black cushion-like protuberance can be observed, showing as moles, i. e. the pseudostroma of the pathogen. The perithecia are subglobose or flat, and clustered in the stroma. The ascospores are unicellular and colorless.

Take the diseased leaf of black scurf of bamboo to make hardworked sections, and then observe the morphological characteristics of perithecium, asci and ascospore of *Phyllachora* under a microscope. Draw the morphological illustrations as you observe.

【Results and Discussion】

1. Summarize the symptom characteristics of all provided plant disease specimens, and grasp the key points of diagnosis of plant diseases caused by the fungi in Sphaeriales.

2. Draw the microscopic morphological illustrations of perithecia, asci and ascospores of the genera of *Valsa*, *Gibberella*, *Glomerella* and *Phyllachora* respectively, and make notes.

3. Compile a retrieve table to distinguish the important pathogens in Sphaeriales.

实验七　子囊菌重要植物病原物的识别与鉴定(Ⅳ)
——腔菌和盘菌子囊菌重要属形态观察

【概述】

腔菌纲(Loculoascomycetes)真菌的子囊果为子囊座，呈垫状或壳状，产囊体在子座中发育，子囊座中心消解可形成单个或多个子囊腔，最终由残留的子座外壳形成子囊的包被，外观与球壳目真菌的子囊壳类似，因此，也称其为假子囊壳(Pseudoperithecium)(图 2-6)。腔菌的另一个重要特征是子囊为双层壁。与植物病害关系较密切的主要有座囊菌目(Dothideales)、多腔菌目(Myriangiales)和格孢腔菌目(Pleosporales)。

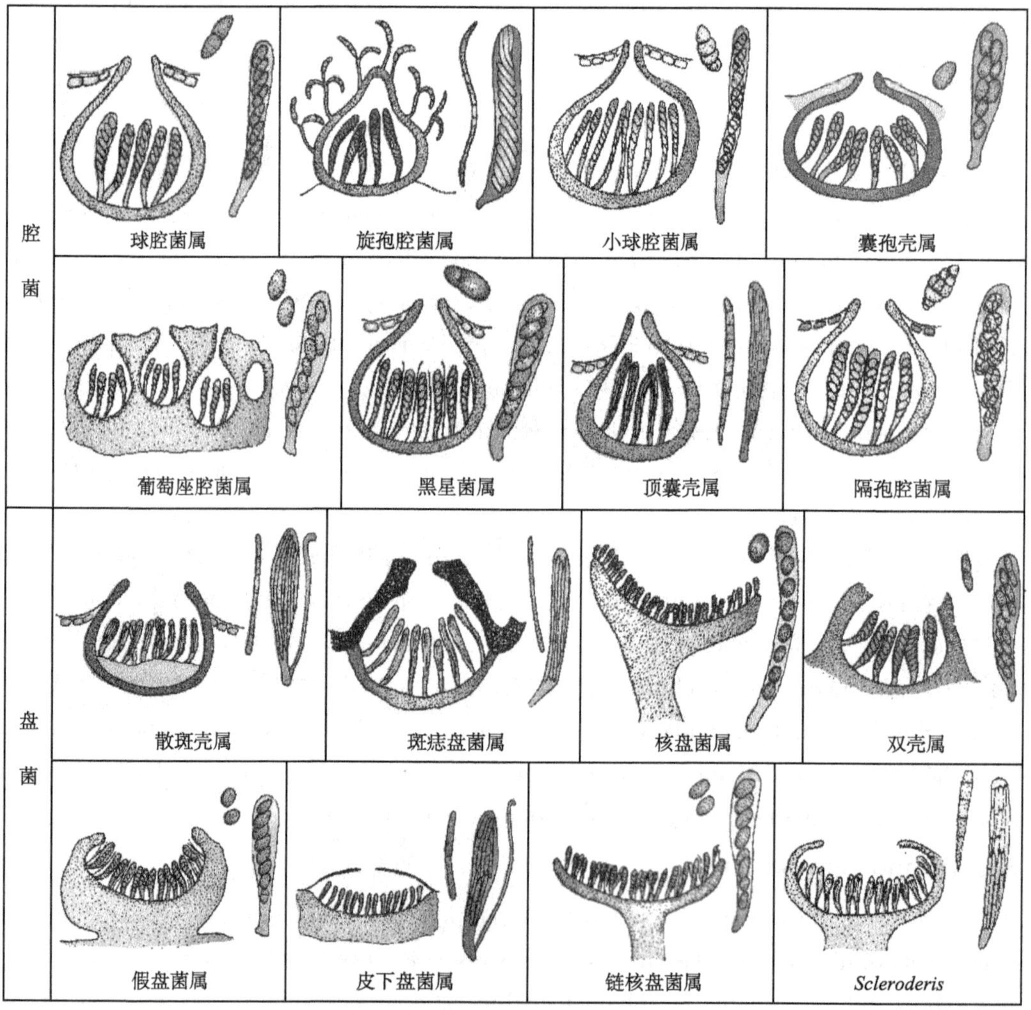

图 2-6　腔菌和盘菌子囊菌重要病原物的形态特征

(引自 Agrios，2005)

盘菌纲(Discomycetes)真菌的子囊果为子囊盘，不同盘菌子囊盘的大小、颜色、质地和结构差别很大(图 2-6)。大多数盘菌缺少无性阶段。与植物病害关系较密切的主要是星裂菌目(Phacidiales)和柔膜菌目(Helotiales)。

【实验目的】

1. 掌握腔菌纲和盘菌纲的子囊菌的重要鉴别依据。

2. 掌握腔菌纲和盘菌纲重要目和代表属所致植物病害症状特点及其显微结构特征。

3. 熟悉徒手切片制作技术。

【材料和器具】

1. 实验材料

下列病原菌引起的植物病害标本或显微玻片。

(1)腔菌纲真菌

①座囊菌目(Dothideales)：杨树灰斑病(*Mycosphaerella mandshurica*)。

②格孢腔菌目(Pleosporales)：梨黑星病(*Venturia pyrina*)、杨树溃疡病(*Botryosphaeria dothidea*)、国槐枝枯病(*Botryosphaeria dothidea*)。

(2)盘菌纲真菌

①星裂菌目(Phacidiales)：槭树漆斑病(*Rhytisma acerinum*)、油松落针病(*Lophodermium conigenum*)。

②柔膜菌目(Helotiales)：油菜菌核病(*Sclerotinia sclerotiorum*)。

2. 实验器具

计算机、投影仪、显微镜、手持放大镜、解剖针、载玻片、盖玻片、通草、刀片、镊子、剪刀、手术刀、吸水纸等。

【方法和步骤】

1. 观察供试植物病害标本的症状，掌握腔菌和盘菌引起植物病害的症状特点。

2. 区分腔菌和盘菌重要属。

①球腔菌属(*Mycosphaerella*)：球腔菌属无性阶段发达，包括 *Pseudocercospora*、*Dothistroma*、*Lecanosticta*、*Septoria*，引起各种植物的叶斑病，常以有性阶段腐生或弱寄生进行越冬。该属假子囊壳散生在寄主叶片表皮下，后期常突破表皮外露，球形或瓶状，子囊圆筒形或棍棒状，子囊孢子无色、双细胞、大小相等。

取杨树灰斑病病叶组织制作徒手切片或直接取该属的永久显微玻片，在显微镜下观察该属的假子囊壳、子囊和子囊孢子的形态特征。边观察边绘制形态图。

②黑星菌属(*Venturia*)：黑星菌属常以无性阶段(*Fusicladium* 或 *Spilocaea*)营寄生生活，可引起叶片、果实及嫩枝的黑星病。该属假子囊壳大多在发病植物残余组织的表皮下形成，黑色、近球形，上部有黑色多隔的刚毛；子囊圆筒形，平行排列于子囊腔内；子囊孢子椭圆形，双细胞，不等大。

取梨黑星病病叶组织制作徒手切片或直接取该属的永久显微玻片，在显微镜下观察该属的假子囊壳、子囊和子囊孢子的形态特征。边观察边绘制形态图。

③葡萄座腔球菌属(*Botryosphaeria*)：该属为兼性寄生菌，引起树木枝干溃疡。该属子

囊座发达，常埋生在寄主表皮下，后突破表皮外露，黑色，炭质，常含多个假子囊壳，罕单生；假子囊壳初成丛埋生于子囊座内，后期突破子囊座外露；子囊间有假侧丝，子囊内壁顶端凹痕明显；子囊孢子单细胞，无色，椭圆形。

取杨树溃疡病或国槐枝枯病发病枝干的小块树皮，务必选取具有明显黑色颗粒的病组织块，制作徒手切片，在显微镜下观察该属的假子囊壳、子囊和子囊孢子的形态特征。边观察边绘制形态图。

④斑痣盘菌属(*Rhytisma*)：该属真菌常引起阔叶植物的漆斑病。在植物发病部位形成的漆斑较大，漆斑的顶部不规则开裂。该黑色漆斑即为病原菌的假子座，内含多个子囊盘。

取槭树漆斑病病叶组织，制作徒手切片，在显微镜下观察该属的子囊盘、子囊和子囊孢子的形态特征。边观察边绘制形态图。

⑤散斑壳属(*Lophodermium*)：该属真菌常引起针叶植物的落针病。发病针叶落地后，会在病叶上产生许多纵向排列的黑色椭圆形突起，即病菌的假子座，内含1个子囊盘，子囊盘棍棒状；子囊孢子单细胞，无色，线状；有侧丝。

取油松落针病病叶组织，制作徒手切片，在显微镜下观察该属的子囊盘、子囊、子囊孢子和侧丝的形态特征。边观察边绘制形态图。

⑥核盘菌属(*Sclerotinia*)：该属真菌的主要特征是菌丝体可以形成菌核，引起多种植物的菌核病。该属真菌的子囊盘褐色，有长柄，产生在菌核上。

取油菜菌核病发病植物材料，肉眼直接观察其菌核的形态，检查是否有子囊盘产生于菌核上。

【结果和讨论】

1. 分别绘制球腔菌属、黑星菌属和葡萄座腔球菌属的假子囊壳、子囊和子囊孢子的显微形态图，并做标注。

2. 分别绘制斑痣盘菌属和散斑壳属的子囊盘、子囊和子囊孢子的显微形态图，并做标注。

3. 假子囊壳和子囊壳的形成过程有何不同？如何区分子囊壳与假子囊壳？

EXPERIMENT 7 Recognition and Identification of Important Plant Pathogens of Ascomycota(Ⅳ)

——Morphological Observation of Important Genera in Loculoascomycetes and Discomycetes

【Introduction】

The ascocarp of the fungi in Loculoascomycetes is ascostroma, which is cushion-like or shell-like. Ascogonium develops in the stroma, and the center of ascostroma can be dissolved to form a single or multiple locules. Finally, the asci are wrapped by the outer shell of residual stoma, which looks like the perithecium of the fungi in Sphaeriales and is also called as pseudoperithecium(Figure

2-6). Another important feature of the fungi in Loculoascomycetes is that the ascus is double-walled. The major groups associated with plant diseases are Dothideales, Myriangiales and Pleosporales.

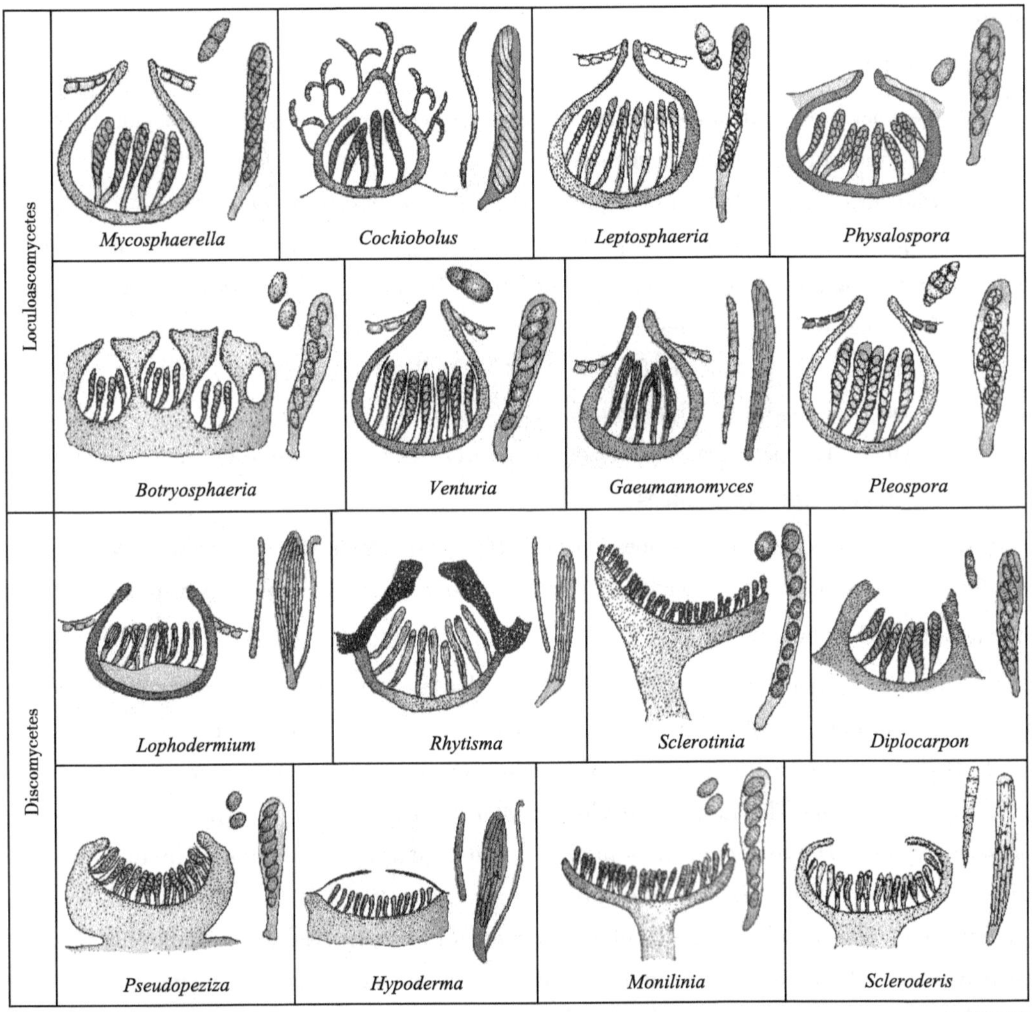

Figure 2-6 Morphological Characteristics of Important Pathogens in Loculoascomycetes and Discomycetes

(Cited from Agrios, 2005)

The ascocarp of the fungi in Discomycetes is apothecium, and the size, color, texture and structure of apothecium vary greatly among different fungi (Figure 2-6). Most fungi in Discomycetes lack an perfect stage. Phacidiales and Helotiales are mainly related to plant diseases.

【Experimental Purpose】

1. Master the important identification basis of the fungi in Loculoascomycetes and Discomycetes.
2. Grasp the symptom characteristics of plant diseases caused by the fungi of the representative

genera of the important orders in Loculoascomycetes and Discomycetes, and master their microscopic structure features.

3. Be familiar with the technique of hand-making section.

【Materials and Apparatus】

1. Materials

The specimens or microslides of plant diseases caused by the following pathogens.

(1) The Fungi in Loculoascomycetes

①Dothideales: The pathogen of poplar gray spot(*Mycosphaerella mandshurica*)

②Pleosporales: The pathogen of pear scab(*Venturia pyrina*), the pathogen of poplar canker (*Botryosphaeria dothidea*) and the pathogen of branch blight of *Sophora japonica*(*B. dothidea*).

(2) The Fungi in Discomycetes

①Phacidiales: The pathogen of maple paint spot(*Rhytisma acerinum*) and the pathogen of pine needle cast(*Lophodermium conigenum*).

②Helotiales: The pathogen of cole sclerotiniose(*Sclerotinia sclerotiorum*).

2. Instruments and Appliances

Computers, projectors, microscopes, hand-held magnifying glasses, dissecting needles, slides, coverslips, ricepaperplant piths, blades, forceps, scissors, scalpels, absorbent papers, etc.

【Methods and Procedures】

1. Observe the symptoms of all provided plant diseases specimens, and master the symptom characteristics of the plant diseases caused by Loculoascomycetes and Discomycetes.

2. Distinguish the important genera of Loculoascomycetes and Discomycetes.

①*Mycosphaerella*: The fungi of *Mycosphaerella* are developed in asexual stage, including *Pseudocercospora*, *Dothistroma*, *Lecanosticta*, and *Septoria*. It can result in leaf spots of a variety of plants, which is usually saprophytic or weak parasitic by its teleomorph for overwintering. The pseudoperithecia of *Mycosphaerella* are scattered under the epidermis of the host leaves and often break through the epidermis in the late stage, showing a spherical or bottle shape. The ascus is cylindrical or club in shape, and the ascospores are colorless and bicellular with equal size.

Take the diseased leaf tissues of poplar gray spots to make handworked sections, or use the permanent slide of *Mycosphaerella* directly, and then observe the morphological characteristics of pseudoperithecium, asci and ascospore of *Mycosphaerella* under a microscope. Draw the morphological illustrations as you observe.

②*Venturia*: The fungi of *Venturia* often conduct their parasitic lives by the type of anamorph (*Fusicladium* or *Spilocaea*), causing scab of leaves, fruits, and shoots. The pseudoperithecium of *Venturia* is mostly formed under the epidermis of the diseased plant residual tissue, which is black and nearly spherical. There are some black septal setae on the upper part of the pseudoperithecium. The ascus is cylindrical and arranged in parallel in the locules. The ascospores are elliptical, and bicellular with unequal size.

Take the diseased leaf tissues of pear scab to make handworked sections, or use the permanent slide of *Venturia* directly, and then observe the morphological characteristics of pseudoperithecium, asci and ascospore of *Venturia* under a microscope. Draw the morphological illustrations as you observe.

③*Botryosphaeria*: The fungi of *Botryosphaeria* are facultative parasites and usually causes stem canker. The ascostroma of *Botryosphaeria* is well developed and often buried under the epidermis of the host, which can break the epidermis to be exposed outside. The ascostroma is black and carbonaceous, in which there are multiple pseudoperithecia, and rarely single one. The pseudoperithecium is buried in the ascostroma in cluster at the early stage and then break the ascostroma to be exposed at the later stage. There are pseudoparaphyses among the asci, and there is an obvious bugholes on the inner wall of ascus. Ascospore is unicellular, colorless and elliptical in shape.

Take a small piece of diseased bark of poplar stem canker or branch blight of *Sophora japonica*, making sure to select the diseased tissues with obvious black particles, and then make handworked sections to observe the morphological characteristics of pseudoperithecium, asci and ascospore of *Botryosphaeria* under a microscope. Draw the morphological illustrations as you observe.

④*Rhytisma*: The fungi of *Rhytisma* usually cause paint spots of broad-leaved plants. The paint spots formed at the diseased site of plant are usually extremely large, and the top of the paint spots is irregularly cracked. The black paint spot is the pseudostroma of *Rhytisma*, containing multiple apothecia.

Take the diseased leaf of maple paint spot to make handworked sections, and then observe the morphological characteristics of apothecia, asci and ascospore of *Rhytisma* under a microscope. Draw the morphological illustrations as you observe.

⑤*Lophodermium*: The fungi of *Lophodermium* often cause needle-cast of coniferophyte. Many black oval protrusions in longitudinal arrangement can be produced on the diseased needles after the needles fall to the ground, i. e. the pseudostroma of *Lophodermium*, in which there is one apothecium. The apothecium is rod-like in shape, the ascospore is unicellular, colorless and linear shape, and there are paraphyses in the apothecium.

Take the diseased needle of pine needle-cast to make handworked sections, and then observe the morphological characteristics of apothecia, asci, ascospore and paraphyses of *Lophodermium* under a microscope. Draw the morphological illustrations as you observe.

⑥*Sclerotinia*: The fungi of *Sclerotinia* are mainly characterized by the formation of sclerotia through hyphae, which can cause sclerotiniose of a variety of plants. The apothecium of *Sclerotinia* is brown, has a long stalk and is produced on the sclerotia.

Take the diseased tissues of oilseed rape sclerotinose, and then observe the morphology of sclerotium directly by naked eyes, and whether the apothecium is produced on the sclerotium.

【Results and Discussion】

1. Draw the microscopic morphological illustrations of pseudoperithecia, asci and ascospores of the genera of *Mycosphaerella*, *Venturia* and *Botryosphaeria* respectively, and make notes.

2. Draw the microscopic morphological illustrations of apothecia, asci and ascospores of the genera of *Rhytisma* and *Lophodermium* respectively, and make notes.

3. What are the differences between the formation of pseudoperithecium and perithecium? How to distinguish the perithecium and pseudoperithecium?

实验八　担子菌重要植物病原物的识别与鉴定(Ⅰ)
——锈菌发育阶段及其重要属形态观察

【概述】

担子菌(Basidiomycota)是菌物中最高等的类群。该门菌物的共同特征是有性生殖产生担子和担孢子。担孢子多生在担子上，每个担子上一般形成4个担孢子。高等担子菌的担子着生在具有高度组织分化的结构上，形成子实层，这种结构称为担子果。低等担子菌的担子裸生，无担子果。

Ainsworth(1973)的分类系统基于担子果的有无和发育类型，将担子菌分为冬孢菌纲(Teliomycetes)、层菌纲(Hymenomycetes)和腹菌纲(Gasteromycetes)。与植物病害关系最密切的是冬孢菌纲。冬孢菌纲属于低等担子菌，不形成担子果，但是形成分散或成堆的冬孢子；广泛分布于世界各地，是非常重要的一类植物病原菌，可分为锈菌目(Uredinales)和黑粉菌目(Ustilaginales)。

锈菌目(Uredinales)真菌一般称为锈菌，绝大多数是植物的专性寄生菌，引起的植物病害称为锈病，常引起农作物的严重损失。锈菌的特点是担子从外生型冬孢子上产生，并以横向隔膜分为4个细胞，每个细胞产生1个小梗和担孢子。

典型锈菌生活史可依次产生5种类型的孢子，不同孢子阶段的锈菌分别用0、Ⅰ、Ⅱ、Ⅲ和Ⅳ表示。0阶段，形成性孢子器及性孢子；Ⅰ阶段，形成锈孢子器及锈孢子；Ⅱ阶段，形成夏孢子和夏孢子堆；Ⅲ阶段，形成冬孢子和冬孢子堆；Ⅳ阶段，形成担孢子。锈菌目的分科依据是性孢子器解剖结构类型；分属的主要依据是冬孢子的形态，其次是夏孢子和锈孢子器(图2-7)。

【实验目的】

1. 认识锈菌5种孢子类型及其繁殖器官结构。
2. 掌握植物锈病的症状识别要点。
3. 掌握锈菌重要属的形态鉴别特征。
4. 熟悉徒手切片制作技术。

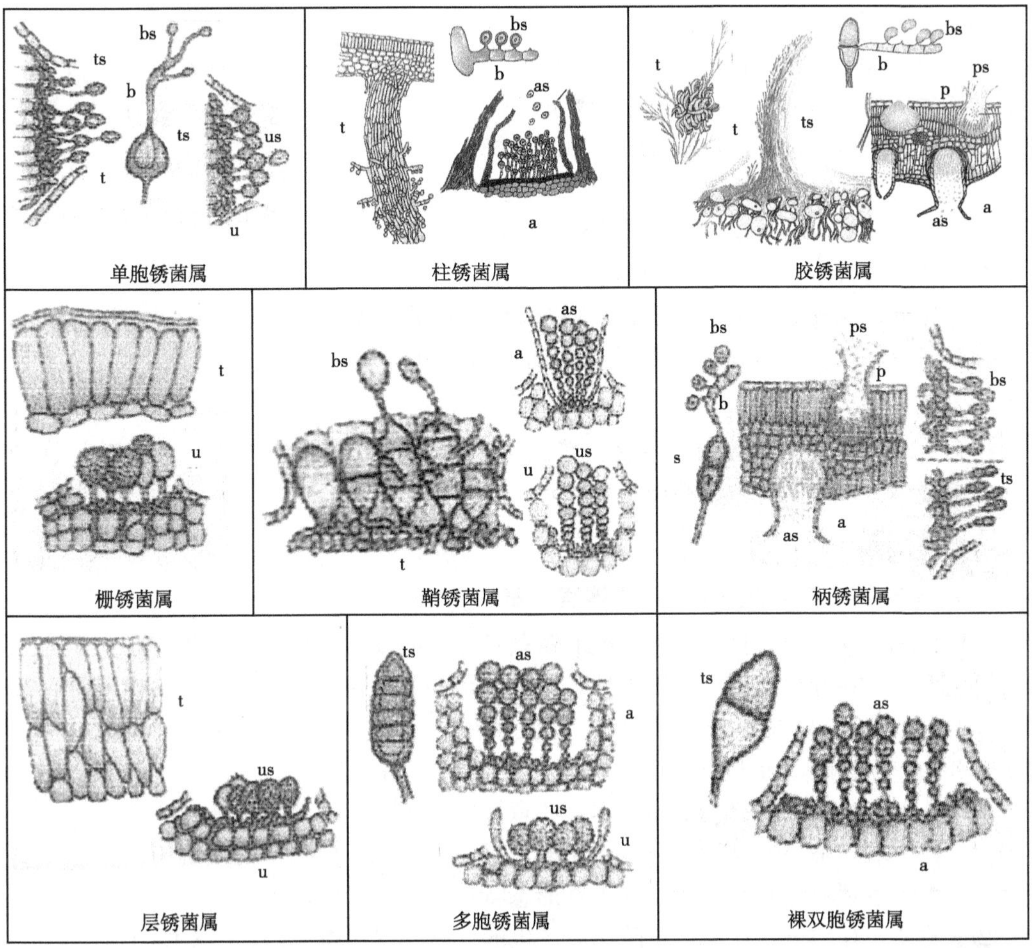

a. 锈孢子器；as. 锈孢子；b. 担子；bs. 担孢子；p. 性子器；ps. 性孢子；t. 冬孢子堆；ts. 冬孢子；u. 夏孢子堆；us. 夏孢子。

图 2-7　锈菌重要病原物的形态特征

（引自 Agrios，2005）

【实验材料和器具】

1. 实验材料

(1)性孢子器及性孢子

苹果、梨、海棠等植物锈病的性孢子器阶段的病害标本，以新鲜材料为宜。

(2)锈孢子器及锈孢子

①锈孢锈属(*Aecidium*)：杯状锈孢子器，如小檗锈病和淫羊藿锈病。

②角孢锈属(*Roestelia*)：毛柱状锈孢子器，如苹果、梨或海棠锈病。

③被孢锈属(*Peridermium*)：疱状或舌状锈孢子器，如松疱锈病和松针锈病。

④裸锈属(*Caeome*)：锈孢子器裸露，如山杏锈病。

(3)夏孢子堆和夏孢子

①柄锈菌属(*Puccinia*)：如小麦锈病，以新发病的小麦叶片为宜。

②栅锈菌属(*Melampsora*)：如杨叶锈病，以新发病的杨树叶片为宜。

(4)冬孢子堆和冬孢子

①单胞锈菌属(*Uromyces*)：如菜豆锈病。

②柄锈菌属(*Puccinia*)：如小麦叶锈病。

③胶锈菌属(*Gymnosporangium*)：如圆柏锈病。

④多胞锈菌属(*Phragmidium*)：如悬钩子锈病和峨眉蔷薇锈病。

⑤栅锈菌属(*Melampsora*)：如落叶松—杨栅锈病。

⑥层锈菌属(*Phakopsora*)：如枣锈病。

⑦柱锈菌属(*Cronartium*)：如锐齿栎锈病和茶藨子锈病。

⑧鞘锈菌属(*Coleosporium*)：如花椒锈病和菊科植物锈病。

2. 实验器具

计算机、投影仪、显微成像系统、普通显微镜、手持放大镜、刀片、载玻片、盖玻片、尖头镊子、通草、剪刀、滴瓶、蒸馏水、吸水纸等。

【方法和步骤】

1. 锈菌各发育阶段的孢子及产孢器官观察

①性孢子及锈孢子：长生活史类型的锈菌冬孢子萌发后产生担子和担孢子，担孢子经传播侵染寄主，在叶片正面产生性子器。性子器最初为橘黄色，并向外分泌出滴液，其中含有大量的性孢子(也称为精子)。受精后性子器变为黑色，而后在正对性子器的叶片背面(少数同时在叶片正面)形成橘黄色的锈孢子器，并产生锈孢子。

观察几种不同类型的锈孢子器标本，注意性子器和锈孢子器在发病组织部位上着生位置的相关性。取新鲜的海棠锈病标本，切取性子器和锈孢子器明显的发病组织，制作徒手切片，显微镜下观察这两种孢子及孢子器形态特点。

②夏孢子：夏孢子是在锈孢子阶段后产生的，常在另一寄主(转主寄主)上形成。大量的夏孢子常聚集在一起形成夏孢子堆，一般初期被埋在寄主表皮下，后期破皮而出，近圆形、椭圆形或矩形，大量粉状夏孢子外露，橙黄色。夏孢子单细胞，球形或椭圆形，橘黄色，多数夏孢子表面有疣刺。

用刀片蘸水轻轻刮取杨树叶锈病的夏孢子堆，制作临时显微玻片，显微镜下观察夏孢子及侧丝的形态。

③冬孢子、担子和担孢子：冬孢子萌发可以形成担子和担孢子。取圆柏锈病的冬孢子堆放置于培养皿中，洒水在25℃保湿24 h后，挑取少许橘黄色的胶状物，制作成临时显微玻片，在显微镜下观察冬孢子、担子和担孢子的形态，并使用显微成像系统拍照记录。

胶锈菌属冬孢子堆形成于寄主表皮下，可突破表皮，红褐色或咖啡色，呈垫状或角状，遇水膨胀胶化后呈橘黄色。冬孢子双细胞，壁薄，具有长柄，柄遇水胶化。

2. 锈菌重要属的区分

①单胞锈菌属(*Uromyces*)：单胞锈菌属大多寄生于豆科植物，单主寄生或转主寄生。冬孢子堆深褐色，冬孢子单细胞，近球形、卵形至椭圆形，有无色短柄，顶端有乳头状突起。

取菜豆锈病发病叶片，先用手持放大镜观察其冬孢子堆有无包被，然后用刀片轻轻刮取冬孢子堆，制作临时显微玻片，在显微镜下观察冬孢子形态特征。边观察边绘制形态图。

②柄锈菌属(*Puccinia*)：柄锈菌属多引起禾本科植物锈病。冬孢子堆产生在寄主表皮下，大多数突破表皮。冬孢子双细胞，壁厚，有短柄，遇水不胶化。注意与胶锈菌属进行区分。

取小麦叶锈病发病叶片，用刀片轻轻刮取冬孢子堆，制作临时显微玻片，在显微镜下观察冬孢子形态特征。边观察边绘制形态图。

③多胞锈菌属(*Phragmidium*)：多胞锈菌属全部单主寄生于蔷薇科植物上。冬孢子堆产生于叶背面，黑色，冬孢子多细胞，深褐色，表面密生疣刺，具不脱落柄，柄上部黄褐色，下部无色、膨大。

取峨眉蔷薇锈病或悬钩子锈病发病叶片，用刀片轻轻刮取冬孢子堆，制作临时显微玻片，在显微镜下观察冬孢子形态特征。边观察边绘制形态图。

④栅锈菌属(*Melampsora*)：栅锈菌属多为转主寄生，0和Ⅰ阶段在落叶松上，Ⅱ和Ⅲ阶段在被子植物上，以杨柳科植物为主。夏孢子堆橘黄色，周围长有许多无色头状侧丝，夏孢子单细胞，表生疣刺；冬孢子堆深褐色，冬孢子单细胞，柱形，无柄，单层紧密排列成栅栏状。

取杨锈病发病叶片，选取夏孢子堆和冬孢子堆明显的发病组织，以通草为固定媒介，制作徒手切片，用显微镜观察夏孢子堆、夏孢子、头状侧丝、冬孢子及其排列特点。边观察边绘制形态图。

⑤层锈菌属(*Phakopsora*)：层锈菌属冬孢子堆生于叶背表皮下，不外露，扁平，深红褐色。冬孢子单细胞，无柄，椭圆形，冬孢子常紧密排列成数层。

取枣锈病发病叶片，选取密生冬孢子堆的发病部位，以通草为固定媒介，制作徒手切片，在显微镜下观察冬孢子形态。边观察边绘制形态图。

⑥柱锈菌属(*Cronartium*)：柱锈菌属全为转主寄生，0和Ⅰ阶段在松属植物上，锈孢子器呈疱状；Ⅱ和Ⅲ阶段在其他植物上。冬孢子单细胞，无柄，柱状，褐色，冬孢子常缠绕在一起形成绳索状的冬孢子柱突破寄主表皮外露。

取锐齿栎锈病或茶藨子锈病的病叶，首先用手持放大镜观察冬孢子柱形态特征，然后用尖头镊子夹取冬孢子柱，制作临时显微玻片，在显微镜下观察冬孢子柱和冬孢子形态特点。

⑦鞘锈菌属(*Coleosporium*)：鞘锈菌属多为转主寄生，0和Ⅰ阶段在松针上，锈孢子器破皮而出，呈显著的舌状包被；Ⅱ和Ⅲ在双子叶植物上，以菊科居多。冬孢子堆蜡质，橘黄色；冬孢子单细胞，无柄，柱状，常排列成单层或多层栅栏状，冬孢子层顶部有透明的胶质鞘。注意与栅锈菌属区分。

取花椒锈病或菊科植物锈病发病叶片，选取密生冬孢子堆的发病部位，以通草为固定媒介，制作徒手切片，显微镜下观察冬孢子形态。边观察边绘制形态图。

【结果和讨论】

1. 使用显微成像系统拍摄胶锈菌属冬孢子、担子和担孢子形态图，彩色打印后提交。

2. 分别绘制单胞锈菌属、柄锈菌属、多胞锈菌属、栅锈菌属和鞘锈菌属的冬孢子形态图。

3. 编制一检索表，区分锈菌重要病原物属。
4. 如何理解单主寄生和转主寄生？

EXPERIMENT 8 Recognition and Identification of Important Plant Pathogens of Basidiomycota(Ⅰ)
——Observation of Development Stages and Morphology of Important Genera in Uredinales

【Introduction】

Basidiomycota is the highest group of fungi, which is characterized by the formation of basidium and basidiospore for sexual reproduction. Basidiospores are produced on the basidium, and four basidiospores are usually formed on each basidium. The basidia of the higher Basidiomycota are born on highly differentiated structures, forming the fruiting layer, which is called basidiocarp. For the lower Basidiomycota, the hasadia are naked without basidiocarp.

According to the taxonomy system of Ainsworth (1973), Basidiomycota is classified into Teliomycetes, Hymenomycetes and Gasteromycetes based on the presence and developmental types of the basidiocarp. Teliomycetes has the closest relationship with plant diseases, which belongs to the lower Basidiomycota and can't form basidiocarp but scattered or heaped teliospores. Teliomycetes is widely distributed all over the world, which is a very important plant pathogenic group and can be divided into Uredinales and Ustilaginales.

The fungi of Uredinales are generally called rust fungi, most of which are obligate parasites in plants. The plant diseases caused by Uredinales are called rust, which often cause serious losses of crops. The primary characteristic of rust fungi is that the basidium is produced from the exogenous teliospores and divided into four cells by the transverse septum, and each cell produces a sterigma and basidiospore.

The rust fungi with typical life history can produce five types of spores, and the rust fungi are expressed as the symbols of 0, Ⅰ, Ⅱ, Ⅲ and Ⅳ at different spore stages. The stage when pycniospores and pycnia are formed is expressed as 0. The stage when aeciospores and aecia are formed is expressed as Ⅰ. The stage when urediospores and uredia are formed is expressed as Ⅱ. The stage when teliospores and telia are formed is expressed as Ⅲ. The stage when basidiospores formed is expressed as Ⅳ. The classification of Uredinales on the level of family is based on the anatomical structure of pycnium, while on the level of genera it is the morphology of teliospore, followed by the urediospore and aecium(Figure 2-7).

【Experimental Purpose】

1. Recognize the five types of spores and reproductive organ structure of rust fungi.
2. Master the key points of symptom recognition of plant rust.

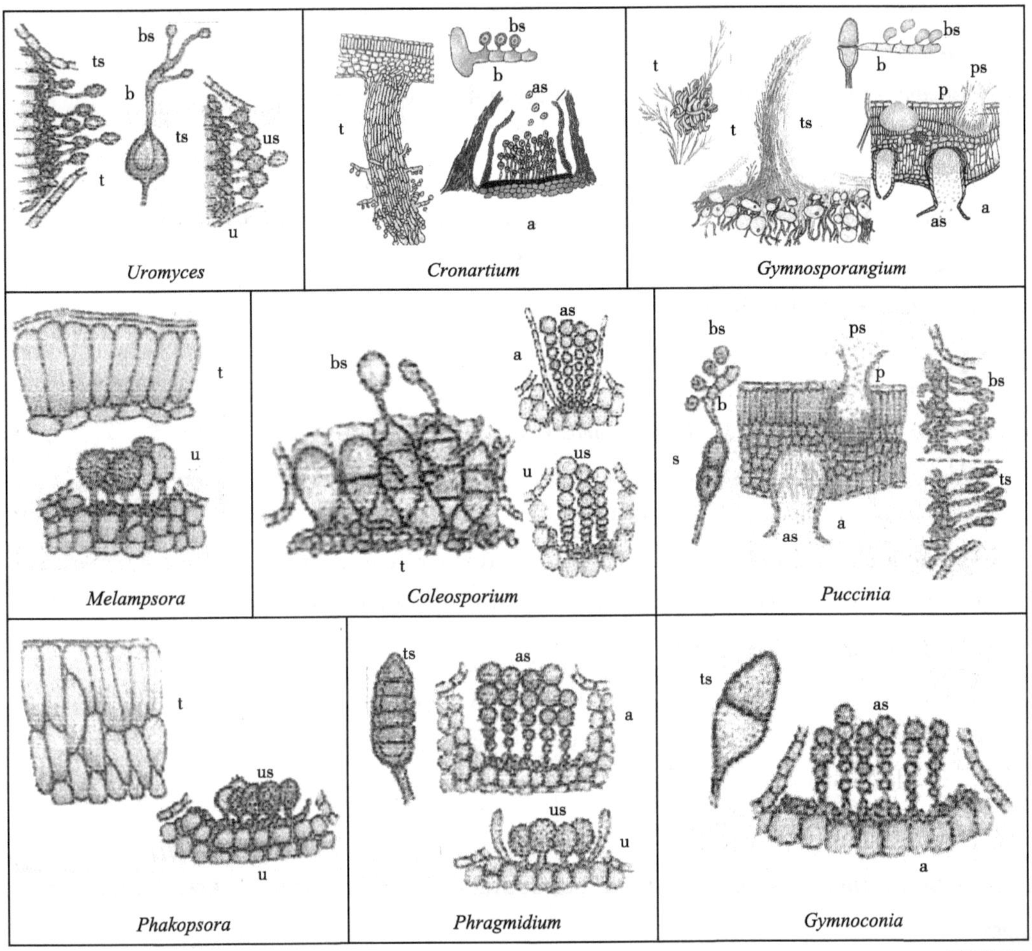

a. aecium; as. aeciospore; b. basidium; bs. basidiospore; p. pycnium; ps. pycniospore;
t. telium; ts. teliospore; u. uredium; us. urediospore.

Figure 2-7　Morphological Characteristics of Important Rust Fungi

(Cited from Agrios, 2005)

3. Master the morphological identification features of important genera of rust fungi.

4. Be familiar with the technique of hand-making section.

【Materials and Apparatus】

1. Materials

(1) Pycnium and Pycniospores

Please provide the specimens of apple rust, pear rust and crabapple rust at the pycnium stage, which are preferable using fresh materials.

(2) Aecium and Aeciospores

①*Aecidium*: The aecium is cup-shaped, such as barberry rust and epimedium rust.

② *Roestelia*: The aecium is hairy column-shaped, such as apple rust, pear rust and crabapple rust.

③*Peridermium*: The aecium is blister-or tongue-shaped, such as pine blister rust and needle rust.

④*Caeome*: The aecium is naked, such as wild apricot rust.

(3) Uredium and urediospores

①*Puccinia*: Wheat leaf rust, which is preferable using the fresh diseased leaves.

②*Melampsora*: Poplar leaf rust, which is preferable using the fresh diseased leaves.

(4) Telium and teliospores

①*Uromyces*: Such as bean rust.

②*Puccinia*: Such as wheat leaf rust.

③*Gymnosporangium*: Such as *Sabina chinensis* rust.

④*Phragmidium*: Such as *Rosa omeiensis* rust and raspberry rust.

⑤*Melampsora*: Such as larch-poplar rust.

⑥*Phakopsora*: Such as jujube rust.

⑦*Cronartium*: Such as *Quercus acutidentata* rust and currant rust.

⑧*Coleosporium*: Such as prickly ash rust and rust of Compositae plants.

2. Instruments and Appliances

Computers, projectors, microscopic imaging systems, microscopes, hand-held magnifying glasses, blades, slides, coverslips, pointed tweezers, ricepaperplant piths, scissors, drop bottles, distilled water, blotting papers, etc.

【Methods and Procedures】

1. Observation of Spores and Fruiting bodies of Rust Fungi at Different Developmental Stages

①Pycniospore and aeciospore: The basidia and basidiospores are produced by the germination of teliospores for the rust fungi with macrocyclic life cycle. The basidiospores are spread to infect host plants, and then the pycnia are produced on the front side of leaf. The pycnium is initially orange-yellow and secretes drips that contain a large number of pycniospores (also known as sperm). The pycnium becomes black after fertilization, and then, the orange-yellow aecium is formed on the back side of the leaf, directly opposite the pycnium, (in a few cases, also on the front side of leaf) and produce aeciospores.

Observe the different types of aecia, and pay attention to the correlation of the location of pycnia and aecia on the diseased tissues. Take the fresh specimen of crabapple rust, choose the diseased tissues with obvious pycnia and aecia to make handworked sections, and then observe the morphological characteristics of the above spores, pycnia and aecia under a microscope.

②Urediospore: Urediospores are formed after the stage of aeciospore, which are usually produced on another host (alternative host). A large number of urediospores often gather together to form uredia. The uredia are usually buried under the host epidermis in the early stage, which are exposed out of the epidermis in the late stage. The uredium is nearly round, oval or rectangular, on the surface of which there is a large number of orange yellow powdery urediospores. Urediospore is unicellular, spherical or oval, orange and most of urediospores have spikes on their surface.

Scrap the uredia on the diseased leaf of poplar rust using a blade that is dipped in water before use to make temporary microscopic slides, and then observe the morphological characteristics of urediospore and paraphysis under a microscope.

③ Teliospore, basidium and basidiospore: Teliospores germinate to form basidia and basidiospores. Take the telia of *Sabina chinensis* rust and place them in a Petri dish, which is then sprayed with water and kept at 25℃ for 24 h. Pick up the orange jelly to make temporary microscopic slides, observe the morphological characteristics of teliospores, basidia and basidiospores under a microscope, and take photos using the microscopic imaging system.

The telia of *Gymnosporangium* are formed under the epidermis of host, which can also break out of the epidermis and show reddish brown or brown. The telia are cushion-like or horn-like in shape, which can swell when encountering water to become colloidal with orange-yellow colour. The teliospore is bicellular with a thin wall and has a long stalk. The stalk of teliospore will become gelation when encountering water.

2. Differentiation of Important Genera of Rust Fungi

①*Uromyces*: The fungi of *Uromyces* are mostly parasitic in leguminous plants, autoecism or heteroecism. The telia are dark brown. The teliospore is unicellular, nearly spherical oval to oval in shape and has a colorless short stalk, which is also characterized by a papillary structure on its top.

Take the diseased leaf of kidney bean rust, observe whether there are envelop wrapping the telia firstly using a hand magnifying glass, then scrape the telia gently with a blade to make temporary microslides, and observe the morphological characteristics of teliospores under a microscope. Draw the morphological illustrations as you observe.

②*Puccinia*: The fungi of *Puccinia* usually cause rust of gramineous plants. The telia are produced under the host epidermis, most of which break through the epidermis. The teliospore is bicellular with a thick wall and has a short stalk which does not gelatinize when encountering water. Note to distinguish it from the genus of *Gymnosporangium*.

Take the diseased leaf of wheat leaf rust, scrape the telia gently with a blade to make temporary microslides, and then observe the morphological characteristics of teliospores under a microscope. Draw the morphological illustrations as you observe.

③*Phragmidium*: All the fungi of *Phragmidium*, which are autoecism, parasitize in the rosaceous plants. The black telia are produced on the back side of leaf. The teliospore is multicellular, dark brown, has many spikes on its surface and has a stalk that does not fall off. The upper part of stalk is yellow-brown, while the lower part is colorless and swollen.

Take the diseased leaf of *Rosa omeiensis* rust or raspberry rust, scrape the telia gently with a blade to make temporary microslides, and then observe the morphological characteristics of teliospores under a microscope. Draw the morphological illustrations as you observe.

④*Melampsora*: Most fungi of *Melampsora* are heteroecious. The stages of 0 and Ⅰ are parasitic in larch, while the stages of Ⅱ and Ⅲ are in angiosperms, mainly in Salicaceae. The uredia are orange-yellow, surrounded by many colorless capitate paraphyses. Urediospore is unicellular with many spikes on its surface. Telia are dark brown, and the teliospores are unicellular, columnar and sessile, which are closely arranged into a monolayer like a palisade.

Take the diseased leaf of poplar rust, choose the diseased tissues with obvious uredia and telia to make handworked sections using the pith as fixing medium, and then observe the morphological characteristics of uredia, urediospores, capitate proliferation and teliospores under a microscope. Draw the morphological illustrations as you observe.

⑤*Phakopsora*: The telia of *Phakopsora* are formed under the epidermis of the back side of leaves, which are not exposed, flat, dark reddish brown. Teliospores are unicellular, sessile, oval, and often closely arranged into several layers.

Take the diseased leaf of jujube rust, choose the diseased sites with abundant telia to make handworked sections using the pith as fixing medium, and then observe the morphological characteristics of teliospores under a microscope. Draw the morphological illustrations as you observe.

⑥*Cronartium*: All the fungi of *Cronartium* are heteroecism. The stages of 0 and Ⅰ of *Cronartium* are parasitic in *Pinus* with the blister-like aecia, while the stages of Ⅱ and Ⅲ are in other plants. The teliospores are unicellular, sessile, columnar, brown, and often entwined together to form a cordlike column that breaks through the epidermis of the host.

Take the diseased leaf of pine-oak rust or currant rust, observe the morphological features of teliospore column firstly using a hand magnifying glass, then pick up the teliospore column with a pointed tweezer to make temporary microslides, and observe the morphological characteristics of teliospore column and teliospore under a microscope.

⑦*Coleosporium*: Most of the fungi of *Coleosporium* are heteroecism. The stages of 0 and Ⅰ of *Coleosporium* are parasitic in pine needles with the aecia breaking out of the epidermis like a prominent tongue envelope, while the stages of Ⅱ and Ⅲ are parasitic in dicotyledonous plants, especially the compositae plants. The telia are waxy and orange. The teliospore is unicellular, sessile, and columnar, which are often arranged in monolayer or multilayers like palisades. There is a transparent colloidal sheath on the top of the teliospore layers. Note to distinguish it from the genus of *Melampsora*.

Take the diseased leaf of prickly ash rust or Compositae rust, choose the diseased sites with abundant telia to make handworked sections using the pith as fixing medium, and then observe the morphological characteristics of teliospores under a microscope. Draw the morphological illustrations as you observe.

【Results and Discussion】

1. Take photos for the teliospores, basidium and basidiospores of *Gymnosporangium* using a

microscopic imaging system, and print them in color before submission.

2. Draw the microscopic morphological illustrations ofteliospores of *Uromyces*, *Puccinia*, *Phragmidium*, *Melampsora* and *Coleosporium*.

3. Compile aretrieve table to distinguish the important pathogenic genera of rust fungi.

4. How to understand autoecism and heteroecism?

实验九　担子菌重要植物病原物的识别与鉴定(Ⅱ)
——黑粉菌重要属形态观察

【概述】

黑粉菌目(Ustilaginales)真菌是冬孢菌纲中的另外一类重要的植物病原菌，常将黑粉菌目的真菌称为黑粉菌，其特征是形成黑色粉状的冬孢子(也称厚垣孢子)。其与锈菌的主要区别是，黑粉菌的冬孢子是从双核菌丝的中间细胞形成的，担孢子直接着生在先菌丝(不形成小梗)的侧面或顶部，成熟后不能弹射。此外，黑粉菌不是专性寄生菌，大多数为兼性寄生，寄生性较强。

黑粉菌主要危害种子植物，在禾本科和莎草科植物上危害较多。其引起的植物病害常称为黑粉病，典型症状为在植物发病部位形成黑色粉状的冬孢子堆。

黑粉菌的分类主要根据冬孢子的性状，如孢子的大小、形状、纹饰、萌发的方式、孢子堆的形态及是否有不孕细胞等(图 2-8)。但是，许多黑粉菌的种很难从冬孢子的性状区别，所以寄主范围也可以作为种的鉴别依据，这一点与锈菌的分类是相同的。

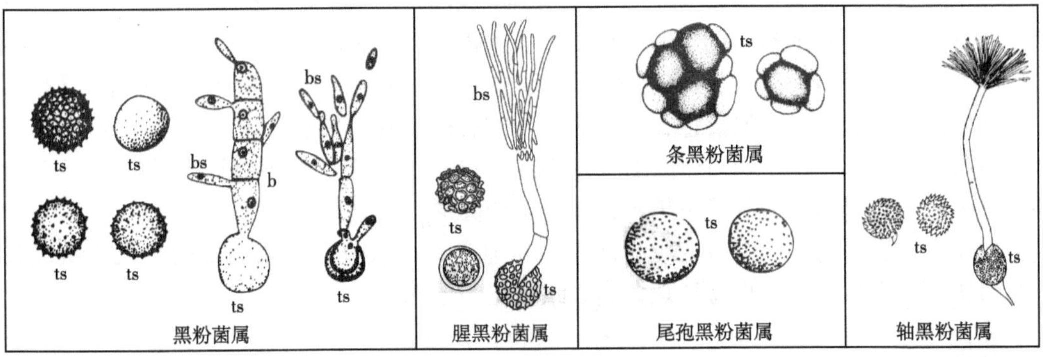

ts. 冬孢子；b. 担子；bs. 担孢子。

图 2-8　黑粉菌重要病原物的形态特征

(引自许志刚，2009)

【实验目的】

1. 掌握植物黑粉病的症状识别特征。
2. 掌握黑粉菌重要属的鉴别特征。
3. 熟悉临时显微玻片制作技术。

【实验材料和器具】

1. 实验材料

植物黑粉病病害标本：小麦散黑穗病（*Ustilago tritici*）、玉米瘤黑粉病（*Ustilago maydis*）、小麦矮腥黑穗病（*Tilletia caries* or *T. laevis*）、小麦秆黑粉病（*Urocystis tritici*）、稻粒黑粉病（*Neovossia horrida*）和玉米丝黑穗病（*Sphacelotheca reiliana*）等。此外，还需提供上述植物黑粉病病原物的永久显微切片。

2. 实验器具

计算机、投影仪、普通显微镜、手持放大镜、刀片、载玻片、盖玻片、镊子、剪刀、滴瓶、蒸馏水、吸水纸等。

【方法和步骤】

1. 观察供试植物黑粉病标本的症状，详细记录其症状特征并加以区分。

2. 根据显微观察，区分黑粉菌的重要病原物。

①黑粉菌属（*Ustilago*）：黑粉菌属冬孢子堆可产生在寄主的各个部位，常在花器，成熟后呈粉状，无菌膜包被，多数黑褐色至黑色，少数浅色，如黄色、紫色等。冬孢子散生，单细胞，表面光滑或有纹饰。

取小麦散黑穗病或玉米瘤黑粉病发病部位的少许黑粉制作临时显微玻片或直接采用病菌的永久显微玻片，在显微镜下观察冬孢子的形态特点。边观察边绘制形态图。

②腥黑粉菌属（*Tilletia*）：腥黑粉菌属冬孢子堆多数产生于寄主子房内，少数产生于寄主的营养器官上，成熟后呈粉状，常有恶腥气味。冬孢子单细胞，外围有无色或淡色的胶质鞘，表面常有网状或刺状突起。

取小麦矮腥黑穗病发病部位的黑粉制作临时显微玻片或直接采用病菌的永久显微玻片，在显微镜下观察冬孢子的形态特点。边观察边绘制形态图。

③条黑粉菌属（*Urocystis*）：条黑粉菌属冬孢子堆生于叶、叶鞘及秆的表皮下，形成条纹，初期埋生于表皮下，铅黑色，成熟后常突破表皮，露出黑色粉末。可形成孢子球，孢子圆球形至椭圆形，由1~3个孢子组成，偶有4个，外围有一层不孕细胞层包被，不孕细胞几乎无色至淡褐色。冬孢子单细胞，圆形至卵圆形，壁光滑。

取小麦秆黑粉病发病部位的黑粉制作临时显微玻片或直接采用病菌的永久显微玻片，在显微镜下观察冬孢子的形态特点。边观察边绘制形态图。

④尾孢黑粉菌属（*Neovossia*）：尾孢黑粉菌属冬孢子堆产生于寄主子房内，为颖壳包被，破坏部分小穗，或仅危害种子的一部分，产生黑粉，半黏结至粉末状。冬孢子单生于产孢菌丝末端细胞内，孢子形成后菌丝残留物在孢子外形成一柄状结构。冬孢子单细胞，球形、广卵形或椭圆形，表面布满齿状突起，突起无色或几乎无色。

取稻粒黑粉病发病部位的黑粉制作临时显微玻片或直接采用病菌的永久显微玻片，在显微镜下观察冬孢子的形态特点。边观察边绘制形态图。

⑤轴黑粉菌属（*Sphacelotheca*）：轴黑粉菌属冬孢子堆生于寄主各部位，以花器居多，团粒状或粉状，初期有菌丝形成的假膜包被，成熟后假膜分离成团状或链状的不孕细胞，

不孕细胞球形、椭圆形或圆柱形，无色或淡色。冬孢子散生，单细胞，孢子堆中有一由寄主残余组织所形成的中轴。

取玉米丝黑穗病发病部位的黑粉制作临时显微玻片或直接用病菌的永久显微玻片，在显微镜下观察冬孢子的形态特点。边观察边绘制形态图。

【结果和讨论】

1. 分别绘制黑粉菌属、腥黑粉菌属、条黑粉菌属、尾孢黑粉菌属和轴黑粉菌属的冬孢子形态图。

2. 编制黑粉菌重要属的检索表。

3. 谈谈锈菌和黑粉菌的异同。

4. 谈谈专性寄生菌和兼性寄生菌的不同。

EXPERIMENT 9　Recognition and Identification of Important Plant Pathogens of Basidiomycota(Ⅱ)

——Morphological observation of important genera in Ustilaginales

【Introduction】

The fungi of Ustilaginales are another important pathogen group in Teliomycetes, which are usually named smut fungi characterized by the formation of black powdery teliospores(also known as chlamydospore). The main difference between smut fungi and rust fungi is that the teliospore of smut fungi is formed from the intermediate cell of the dikaryotic hypha, and the basidiospore directly grows on the side or top of promycelium without forming small pedicels and can't eject after maturity. Besides, the smut fungi are not obligate, and most of them are facultative with strong parasitism.

The smut fungi mainly do harm to seed plants, especially the plants of Gramineae and Cyperaceae. The plant diseases caused by smut fungi are often called smut and are characterized by the formation of black powdery telium on the diseased tissues.

The classification of smut fungi is mainly carried out according to the features of teliospores, such as the size, shape, ornamentation and germination mode of teliospore, the shape of telium, and whether there are infertile cells or not(Figure 2-8). However, many species of smut fungi are hardly differentiated based on the features of teliospores. Thus, the host range of smut fungi can also be taken as an identification indicator of species, which is the same as that of rust fungi.

【Experimental Purpose】

1. Master the recognition features of symptoms for plant smut.

2. Master the identification characteristics of the important genera of smut fungi.

3. Be familiar with the technique of making temporary microslides.

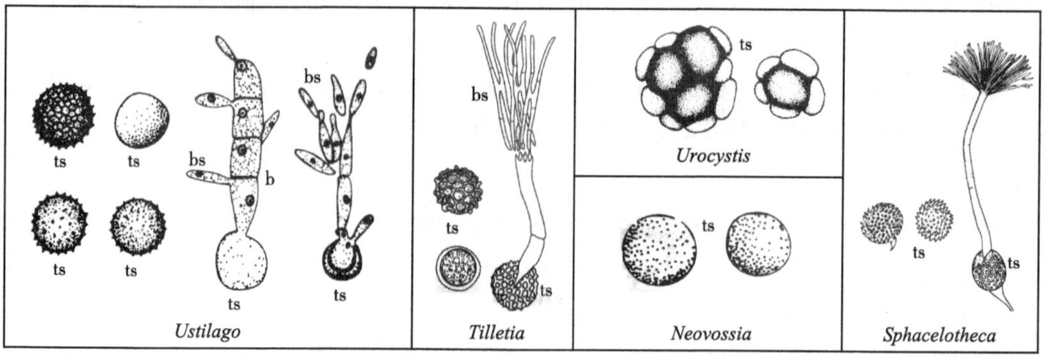

ts. teliospore; b. basidium; bs. basidiospore.

Figure 2-8 Morphological Characteristics of Important Smut Fungi

(Cited from Zhigang Xu, 2009)

【Materials and Apparatus】

1. Materials

The specimens of plant smut: wheat loose smut (*Ustilago tritici*), corn smut (*Ustilago maydis*), wheat dwarf smut (*Tilletia caries*, *T. laevis*), wheat stem smut (*Urocystis tritici*), rice grain smut (*Neovossia horrida*) and corn head smut (*Sphacelotheca reiliana*). Moreover, the permanent slides of the pathogens of the above plant smut should also be provided.

2. Instruments and Appliances

Computers, projectors, microscopes, handheld magnifying glasses, blades, glass slides, coverslips, tweezers, scissors, dropping bottles, distilled water, blotting papers, etc.

【Method and Procedure】

1. Observe the symptoms of all provided specimens of plant smut, record their symptom characteristics and then distinguish them.

2. Distinguish the important pathogens of plant smut based on the microscopic observation.

①*Ustilago*: The telia of *Ustilago* can be produced in various parts of the host, usually in the floral organs. The mature telia are powdery without mycoderm, most of which are black brown to black, and a few are light colored, such as yellow and purple, Teliospores are scattered and unicellular, the surface of which is smooth or ornamented.

Take a little black powder on the diseased tissues of wheat loose smut or corn smut to make temporary microscopic slides, or use the permanent slides of the pathogens, and then observe the morphological features of teliospores under a microscope. Draw the morphological illustrations as you observe.

②*Tilletia*: Most telia of *Tilletia* are produced in the ovary of the host, and just a few are produced in the vegetative organs of the host. The mature telia are powdery with a fishy smell. The teliospores are unicellular and surrounded by a layer of colorless or light colloid sheathe. The surface of teliospore is often netlike or spiny.

Take the black powder on the diseased sites of wheat dwarf smut to make temporary microscopic slides, or use the permanent slides of the pathogen, and then observe the morphological features of teliospores under a microscope. Draw the morphological illustrations as you observe.

③*Urocystis*: The telia of *Urocystis* are formed under the epidermis of leaf sheaths and culms, forming stripes. The telia are buried under the epidermis at the initial stage and are leaden black, while the mature telia can break through the epidermis and expose as black powder. The spore ball can be formed with a round to oval shape, which is composed by 1-3 spores, occasionally four spores. There is a layer of infertile cells wrapping around the spore ball. The infertile cells are almost colorless to light brown. The teliospore is unicellular, round to oval, the wall of which is smooth.

Take the black powder on the diseased sites of wheat stem smut to make temporary microscopic slides, or use the permanent slides of the pathogen, and then observe the morphological features of teliospores under a microscope. Draw the morphological illustrations as you observe.

④*Neovossia*: The telia of *Neovossia* are produced in the ovary of host plant, which is enveloped by the glume and destroys part of the spikelet or only part of the seed, and then the black powder is formed showing the status of half cementation to powder. The teliospore is formed in the terminal cell of the conidiogenous hyphae, and the hyphal residue forms a stalk structure outside the spores after spore formation. The teliospore is unicellular, and spherically, broadly ovate or elliptic in shape, the surface of which is covered with many colorless or almost colorless dentate structures.

Take the black powder on the diseased sites of rice grain smut to make temporary microscopic slides, or use the permanent slides of the pathogen, and then observe the morphological features of teliospores under a microscope. Draw the morphological illustrations as you observe.

⑤*Sphacelotheca*: The telia of *Sphacelotheca* are formed in various parts of the host plant, mostly in flower apparatus, which are granular or powdery. At the initial stage, the telia are coated by the pseudomembrane formed by hyphae, and after maturation, the pseudomembrane is separated into clusters or chains of infertile cells. The infertile cells are spherical, oval or cylindrical, colorless or pale. The teliospores are dispersive, unicellular, and there is a central axis formed by residual host tissue in the telia.

Take the black powder on the diseased sites of corn head smut to make temporary microscopic slides, or use the permanent slides of the pathogen, and then observe the morphological features of teliospores under a microscope. Draw the morphological illustrations as you observe.

【Results and Discussion】

1. Draw the morphological illustrations of the teliospores of *Ustilago*, *Tilletia*, *Urocystis*, *Neovossia* and *Sphacelotheca*, respectively.

2. Compile a retrieve table of the important genera of smut fungi.
3. Talk about the similarities and differences between rust fungi and smut fungi.
4. Talk about the differences between obligate parasites and facultative parasites.

实验十　无性型真菌重要植物病原物的识别与鉴定(I)
——丝孢纲重要属形态观察

【概述】

无性型真菌(Anamorphic fungi)通常指那些只有无性态或有性态尚未被发现的真菌。然而，随着研究的深入，部分无性型真菌的有性态陆续被发现，且大多属于子囊菌，少数属于担子菌。由于无性型真菌包含了许多子囊菌和担子菌的无性态，这就导致了同一个种交叉分在不同的分类单元中，随之带来的问题就是同一个物种就有两个学名。根据《国际藻类、真菌及植物命名法规》，每个生物的种只能有一个正式的学名，对于许多子囊菌和担子菌来说，它们有性态学名是正式的学名，然而，它们的无性态(分生孢子阶段)学名在应用上很方便，所以目前在国际上也认为是合法的，故无性态的学名也为合法学名，主要原因是这些真菌的无性态极其发达，且具有重要经济意义，与人类关系密切。人们在发现它的无性态后先给予一个学名，后来又发现它的有性态，于是又有了有性态的学名，由于它的有性态较少见或不重要，且根据其无性态特征很容易进行分类和鉴定，故人们常习惯使用它的无性态学名。因此，无性态学名仍然被广泛使用，而有性态学名反而较少使用。

无性型真菌在自然界分布广泛，种类繁多，目前主要以其分生孢子发育方式作为基本分类依据。因此，无性型真菌是由许多系统发育关系并不密切的真菌构成，它们只是在产孢形式上相似的类群的聚集。无性型真菌中有许多种类是植物病原菌，所引起的植物病害几乎涵盖了所有病害症状类型，大多数为兼性寄生菌和兼性腐生菌。熟练掌握无性型真菌的鉴定方法，对于植物病原学研究至关重要。

Ainsworth(1973)的分类系统将无性型真菌分为芽孢纲(Blastomycetes)、丝孢纲(Hyphomycetes)和腔孢纲(Coelomycetes)。芽孢纲真菌大多腐生，有些寄生在人和动物体内，与植物病害无关。

丝孢纲真菌大多数是植物的重要病原菌，其主要特点是分生孢子梗散生、束生或着生在分生孢子座上，梗上着生分生孢子。此外，有些种类不产生分生孢子，只形成厚垣孢子(图 2-9)。丝孢纲分为无孢目(Agonomycetales)、丝孢目(Hyphomycetales)、束梗孢目(Stilbellales)和瘤座孢目(Tuberculariales)。与植物病害关系比较密切的是无孢目、丝孢目和瘤座孢目。

【实验目的】

1. 了解无性型真菌的分类依据，掌握丝孢纲真菌的重要鉴定特征。
2. 通过了解丝孢纲重要病原菌的形态特征，掌握其鉴定依据。
3. 熟悉临时显微玻片制作技术。

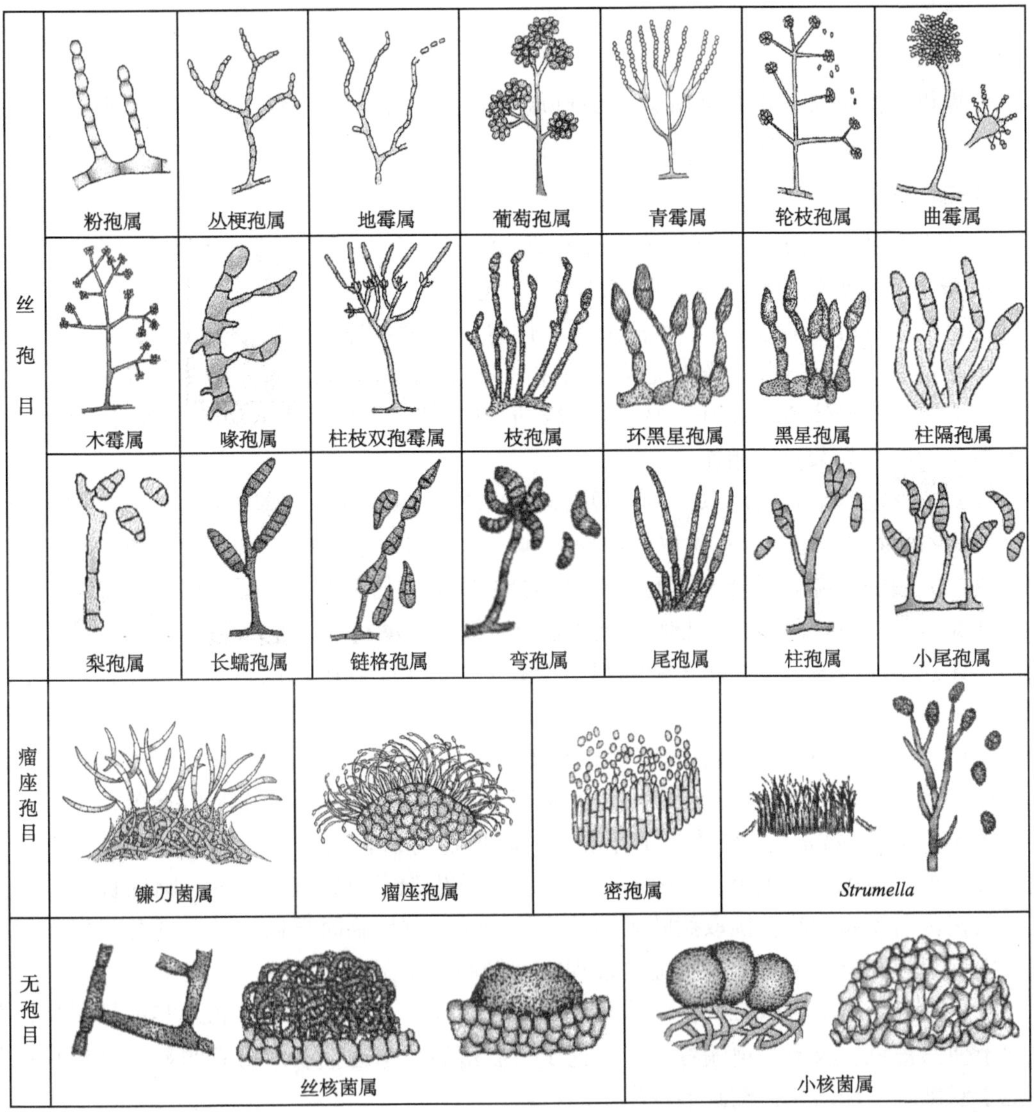

图 2-9　丝孢纲重要植物病原物的形态特征

（引自 Agrios，2005）

【材料和器具】

1. 实验材料

准备下列病菌的培养物、永久玻片或病菌引起的病害标本。

①无孢目：水稻纹枯病（*Rhizoctonia solani*）和松苗猝倒病（*R. solani*）。

②丝孢目：稻瘟病（*Pyricularia grisea*）、正木白粉病（*Oidium euonymi-japonici*）、杨黑星病（*Fusicladium tremulae*）、芍药褐斑病（*Cladodporium paeoniae*）、玉米大斑病（*Exserohilum turcicum*）、玉米小斑病（*Bipolaris maydis*）、樱花褐斑病（*Cercospora cerasella*）和链格孢属（*Alternaria* spp.）。

③瘤座孢目：花椒根腐病(*Fusarium solani*)。

2. 实验器具

计算机、投影仪、普通显微镜、手持放大镜、单面刀片、双面单片、载玻片、盖玻片、通草、尖头镊子、剪刀、滴瓶、蒸馏水、吸水纸等。

【方法和步骤】

1. 观察所提供的由丝孢纲真菌引起的植物病害的症状，掌握其症状识别要点。

2. 丝孢纲重要病原物的显微形态观察与区分。

①丝核菌属(*Rhizoctonia*)：丝核菌属兼性寄生，主要为害植物根部，引起多种植物的猝倒和立枯病。该属真菌不产生无性繁殖体。菌丝无色至褐色，菌丝多为直角分枝，近分枝处形成隔膜与母体菌丝隔开，呈缢缩状；形成菌核，菌核褐色或黑色，内外颜色一致，着生在菌丝中，与菌丝体相连。

观察丝核菌属培养物形态，注意观察幼龄和老龄菌丝颜色的差异(幼龄无色，老龄褐色)，以及菌核的形态特征；挑取少许菌丝体，制作临时显微玻片，在显微镜下观察菌丝的颜色、分枝处的缢缩、分支附近的隔膜和分枝间形成的角度，绘图描述这些特征。

②梨孢属(*Pyricularia*)：梨孢属分生孢子梗细长，淡褐色，呈屈膝状弯曲；分生孢子梨形，无色至淡橄榄色，大多 3 个细胞，少数 2 个或 4 个细胞。该属真菌寄生性极强，主要危害禾本科植物。

取稻瘟病发病小枝梗或病穗茎(经保湿处理)，挑取病部灰白色霉层，制作临时显微玻片，在显微镜下观察其分生孢子梗和分生孢子形态特征。边观察并绘制形态图。

③粉孢属(*Oidium*)：该属专性寄生，引起植物的白粉病。菌丝体生于寄主表面，白色。分生孢子梗无色，短小，无分隔，不分枝；分生孢子单细胞，无色，圆柱状，串生。

取正木白粉病病叶，挑取病叶表面的少许白粉，制作临时显微玻片，在显微镜下观察其分生孢子梗和分生孢子形态特征。边观察并绘制形态图。

④黑星孢属(*Fusicladium*)：该属真菌寄生性较强，常引起植物叶和果实的黑星病。分生孢子梗黑褐色，较短，顶端着生分生孢子，脱落后分生孢子梗上留有明显的孢子痕；分生孢子双细胞，深褐色，椭圆形至梨形。

取杨树黑星病的发病叶片，选取具有明显黑色霉层的发病部位，以通草进行固定，制作徒手切片，在显微镜下观察其分生孢子梗和分生孢子形态特征。边观察并绘制形态图。

⑤芽枝霉属(*Cladosporium*)：该属真菌弱寄生，引起植物的叶霉病。分生孢子梗黑色、不分枝或近顶部分枝，顶端着生分生孢子，脱落后分生孢子梗上留有明显的孢子痕；分生孢子橄榄褐色，卵形，圆柱形或不规则，1~4 个细胞，常形成孢子链。

取芍药褐斑病的发病叶片，选取具有明显褐色霉层的发病部位，以通草进行固定，制作徒手切片，显微镜下观察其分生孢子梗和分生孢子形态特征。边观察并绘制形态图。

⑥突脐蠕孢属(*Exserohilum*)：分生孢子梗粗壮，黑褐色，顶部合轴式延伸；分生孢子梭形、圆筒形或倒棍棒状，深褐色，脐点突出，多细胞。

取玉米大斑病的发病叶片，选取具有明显褐色霉层的发病部位，以通草进行固定，制作徒手切片，在显微镜下观察其分生孢子梗和分生孢子形态特征。边观察并绘制形态图。

⑦离蠕孢属(*Bipolaris*)：分生孢子梗形态与突脐蠕孢属相似；分生孢子长梭形，直或弯曲，深褐色，多细胞，脐点不突出，位于基细胞内。

取玉米小斑病的发病叶片，选取具有明显褐色霉层的发病部位，以通草进行固定，制作徒手切片，在显微镜下观察其分生孢子梗和分生孢子形态特征。边观察并绘制形态图。注意与突脐蠕孢属进行区分。

⑧尾孢属(*Cercospora*)：该属真菌分生孢子梗褐色，不分枝，丛生于小型子座上或表生于菌丝上，合轴式发育，梗上有明显的孢痕；分生孢子棍棒状至线形，直或弯曲，多细胞，浅色至深褐色。

取樱花褐斑病的发病叶片，选取症状明显的发病部位，以通草进行固定，制作徒手切片，在显微镜下观察其分生孢子梗和分生孢子形态特征。边观察并绘制形态图。

⑨链格孢属(*Alternaria*)：该属真菌腐生或弱寄生，以腐生为主，可危害多种植物，引起叶斑及果实腐烂。分生孢子梗较长，橄榄褐色或暗褐色，多不分枝，合轴式延伸；分生孢子棍棒状、椭圆形或卵圆形，褐色，具有横、纵或斜隔膜，顶端无喙或有喙，单生或串生。

用尖头镊子挑取链格孢属培养物少许，制作临时显微玻片，在显微镜下观察其分生孢子梗和分生孢子形态特征。边观察并绘制形态图。

⑩镰孢属(*Fusarium*)：镰孢属是瘤座孢目的代表，特点是分生孢子梗着生在分生孢子座上。分生孢子梗形状不一，不分枝，聚集形成垫状的分生孢子座。分生孢子无色，有两种类型，大型分生孢子多细胞、镰刀形，小型分生孢子单细胞，椭圆形或圆形。有的镰刀菌在逆境胁迫时，还会产生厚垣孢子。

取花椒根腐病菌培养物少许，制作临时显微玻片，在显微镜下观察其分生孢子座、分生孢子梗和分生孢子形态特征。边观察并绘制形态图。

【结果和讨论】

1. 总结供试植物病害标本的症状识别特点。

2. 分别绘制梨孢属、粉孢属、黑星孢属、芽枝霉属、突脐蠕孢属、离蠕孢属、尾孢属、链格孢属和镰孢属的分生孢子梗和分生孢子形态图。

3. 编制本实验中涉及的无性型真菌重要属的检索表。

4. 谈谈无性型真菌与子囊菌之间的关系。

EXPERIMENT 10　Recognition and Identification of Important Plant Pathogens of Anamorphic Fungi(Ⅰ)

—— Morphological Observation of Important Genera in Hyphomycetes

【Introduction】

Anamorphic fungi usually refers to those fungi that just have anamorph or their teleomorph has not been discovered. However, as the deepening of research, the teleomorph of some anamorphic fungi has been successively discovered, most of which belong to Ascomycota and a few belong to Basidiomycota. Due to the fact that the anamorph of many fungi in Ascomycota and Basidiomycota are contained in anamorphic fungi, it results in that the same species is crossly divided in different taxa, the problem accompanied with which is that there are two scientific names for the same species. According to *International Code of Nomendature for Algae, Fungi, and Plant*, each creature can only have a kind of formal scientific name. For many fungi in Ascomycota and Basidiomycota, their scientific names of teleomorph are the formal names. However, the application of their scientific names of anamorph, i. e. the conidiospore stage, is very convenient, which is also considered to be legitimate at present in the world. Therefore, the scientific names of anamorph are also the legitimate names, the main reason of which is that the anamorph of these fungi are highly developed, economically important and closely related to humans. A scientific name is firstly given to the fungus after its anamorph is discovered by people, while it will have a scientific name of teleomorph when its teleomorph is discovered. Because the teleomorph of these fungi is rarely observed or not important, and it is easy to classify and identify them based on their anamorph characteristics, people are used to use the scientific names of anamorph. Hence, the scientific names of anamorph are still widely applied, while the scientific names of teleomorph are less used.

Anamorphic fungi are widely distributed in nature with many species, which are currently classify based on the development model of conidiospores. Therefore, anamorphic fungi are composed by the fungi with no close phylogenetic relationships, which are only the aggregation of similar groups in sporulation form. Many anamorphic fungi are plant pathogens, and the plant diseases caused by them cover almost all symptom types, most of which are facultative parasites and facultative saprophytes. In practice, it is very important to master the identification method of anamorphic fungi in plant etiology.

According to the classification system of Ainsworth (1973), anamorphic fungi are classified into Blastomycetes, Hyphomycetes, and Coelomycetes. Most Blastomycetes are saprophytic, some are parasitic in humans and animals and not associated with plant diseases.

Most of the fungi in Hyphomycetes are important plant pathogens, the main characteristics of which are that the conidiophores are scattered, bundled or borne on the sporodochium, and conidia are produced on the conidiophores (Figure 2-9). In addition, some species do not produce conidia but only chlamydospores. Hyphomycetes can be divided into Agonomycetales, Hyphomycetales, Stilbellales, and Tuberculariales, of which Agonomycetales, Hyphomycetales, and Tuberculariales are more closely related with plant diseases.

【Experimental Purpose】

1. Understand the classification basis of fungi imperfecti, and master the important identification characteristics of Hyphomycetes fungi.

2. Master the identification basis of Hyphomycetes by understanding the morphological characteristics of important pathogenic fungi belonging to this class.

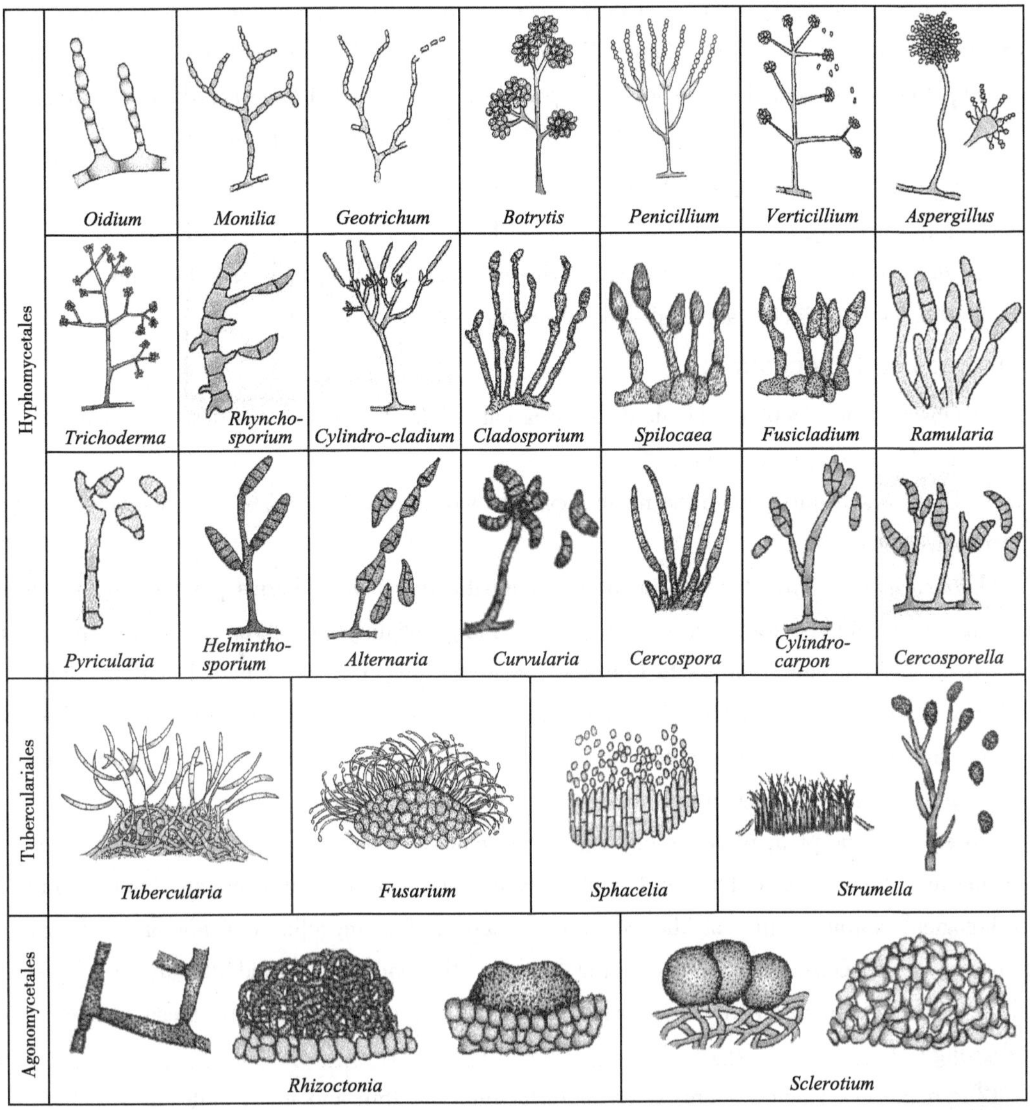

Figure 2-9 Morphologic Characteristics of Important Pathogens in Hyphomycetes

(Cited from Agrios, 2005)

3. Be familiar with the techniques of making temporary microscopeslides.

【Materials and Apparatus】

1. Materials

Prepare permanent slides of cultures of the following pathogens or specimens of diseases caused by the fungi:

①Agonomycetales: Rice sheath blight (*Rhizoctonia solani*) and pine seedling damping off (*R. solani*).

②Hyphomycetales: Rice blast (*Pyricularia grisea*), powdery mildew of *Euonymus japonicus*

(*Oidium euonymi-japonici*), poplar scab(*Fusicladium tremulae*), peony brown spot(*Cladodporium paeoniae*), northern corn leaf blight (*Exserohilum turcicum*), southern leaf blight (*Bipolaris maydis*), Cherry blossoms brown spot(*Cercospora cerasella*), and *Alternaria*(*Alternaria* spp.).

③Tuberculariales: Root rot of *Zanthoxylum bungeanum*(*Fusarium solani*).

2. Instruments and Appliances

Computers, projectors, microscopes, hand-held magnifying glasses, single-sided blades, double-sided blades, slides, coverslips, ricepaperplant piths, pointed tweezers, scissors, dropping bottles, distilled water, absorbent papers, etc.

【Methods and Procedures】

1. Observe the symptoms of plant diseases caused by Hyphomycetes fungi, and master the key points of symptom recognition.

2. Observe and differentiate the microscopic morphological characteristics of important pathogens of Hyphomycetes.

①*Rhizoctonia*: *Rhizoctonia* is facultative parasitie that mainly harms plant roots and causes damping-off and blight of a variety of plants. Fungi of this genus do not produce asexual propagules. Mycelia are colorless to brown. Most of the mycelia are branched at right angles. A diaphragm, which is constricted, is formed near the branches to separate them from the parent mycelia. Sclerotinia are formed, and are brown or black, with same color inside and outside. These sclerotinia are growing in the mycelium, and connected with the mycelia.

Observe the morphology of the culture of the genus *Rhizoctonia*, notice the difference in the color between the young and the old hyphae(young are colorless and old are brown) and the morphological characteristics of the sclerotia. Select a few mycelia and prepare a temporary microscope slide. Observe the color of the hyphae, the constriction and diaphragm at branch, and the angle formed between the branches under the microscope. Describe these characteristics by drawing.

②*Pyricularia*: Conidiophores of *Pyricularia* fungi are thin and long, pale brown, bent-like knee in shape. Conidia are pear-shaped, colorless to pale olive, mostly three-celled, few two-celled or four-celled. The parasitism of the fungi of this genus is extremely strong. They mainly harm Poaceae plants.

Take diseased branchlets or diseased ear stems of rice blast(after moisturizing treatment), pick out the gray-white mold layer of the diseased site to make temporary microscope slides, and then observe the morphological features of conidiophores and conidia under a microscope. Draw the morphological illustrations as you observe.

③*Oidium*: Fungi of the genus are obligate parasitism, causing powdery mildew on plants. The mycelia are white and live on the surface of the host. The conidiophores are colorless, short, non-diaphragmed, and unbranched. The conidia are single celled, colorless, cylindrical, and clustered.

Take diseased leaves of powdery mildew of *E. japonicus*, pick up a little powder to make temporary microscope slides, and then observe the morphological features of conidiophores and conidiospores under a microscope. Draw the morphological illustrations as you observe.

④*Fusicladium*: The fungi of this genus are strong parasitic, which usually cause scab on leaves and fruits. The conidiophores are dark brown and short. Conidia live on the top of the conidiophores. There are obvious spore scars on the conidiophore after shedding. The conidia are double-celled, dark brown, and oval to pear-shaped.

Take the diseased leaf of poplar scab, select the diseased sites with obvious black mould to make hardworked sections using the pith as the fixing medium, and then observe the morphological features of conidiophores and conidiospores under a microscope. Draw the morphological illustrations as you observe.

⑤*Cladosporium*: The fungi of this genus are weakly parasitic and cause leaf mildew in plants. The conidia are black, unbranched or near-apically branched. Conidia live on the top of the conidiophores. There are obvious spore scars on the conidiophore after shedding. Conidia are olive brown, ovoid, cylindrical or irregular, 1-4 celled, usually form spore chains.

Take the diseased leaf of poeny brown spot, select the diseased sites with obvious brown mould to make handworked sections using the pith as the fixing medium, and then observe the morphological features of conidiophores and conidiospores under a microscope. Draw the morphological illustrations as you observe.

⑥*Exserohilum*: The conidiophores are stout, black brown, and coaxial extended at the top; Conidia are fusiform, cylindrical or inverted club-shaped, dark brown, umbilical point protruded, and multicellular.

Take the diseased leaf of northern corn leaf blight, select the diseased sites with obvious brown mould to make handworked sections using the pith as the fixing medium, and then observe the morphological features of conidiophores and conidiospores under a microscope. Draw the morphological illustrations as you observe.

⑦*Bipolaris*: The morphology of conidiophore is similar to that of *Exserohilum*. Conidia are long fusiform, straight or curved, dark brown, multicellular. The umbilical point is not protrusive and inside the basal cell.

Take the diseased leaf of southern corn leaf blight, select the diseased sites with obvious brown mould to make handworked sections using the pith as the fixing medium, and then observe the morphological features of conidiophores and conidiospores under a microscope. Draw the morphological illustrations as you observe. Note to distinguish it from the genus *Exserohilum*.

⑧*Cercospora*: The conidia of this genus fungi are brown and unbranched, clustered on small sporodochium or surface-grown on hyphae, and coaxially developed. There are obvious spore scars on the conidiophore. Conidia are stick to linear, straight or curved, multicellular, light to dark brown.

Take the diseased leaf of cherry blossom brown spot, select the diseased sites with obvious symptom to make handworked sections using the pith as the fixing medium, and then observe the morphological features of conidiophores and conidiospores under a microscope. Draw the morphological illustrations as you observe.

⑨*Alternaria*: The fungi of this genus are saprophytic or weakly parasitic, mainly saprophytic, and can harm a variety of plants, causing leaf spots and fruit decay. The conidiophore is long, olive brown or dark brown, mostly unbranched, coaxial extended. The conidia are clavate shape, obclavate elliptic or ovoid, brown, with transverse, longitudinal or oblique septa, apically beakless or beaked, solitary or in clustered.

Pick a few of the culture of *Alternaria* with pointed tweezers to make temporary microscope slides, and then observe the morphological features of conidiophores and conidiospores under a microscope. Draw the morphological illustrations as you observe.

⑩*Fusarium*: The genus *Fusarium* is a representative of the order Tuberculariales, characterized by conidiophore living on sporodochium. The conidiophores show different shapes. They are unbranched, aggregated to form a padded sporodochium. The conidia are colorless and can be divided into two types. The large conidia are multicellular and sickle; while the small conidia are unicellular, or oval or round. Some *Fusarium* can produce chlamydospores under stress.

Take a little culture of *F. solani* to make temporary microscope slides, and then observe the morphological features of sporodochia, conidiophores and conidia under a microscope. Draw the morphological illustrations as you observe.

【Results and Discussion】

1. Summarize the symptom recognition characteristics of the provided plant disease specimens.
2. Draw the morphological illustrations of the conidiophores and conidia of *Pyricularia*, *Oidium*, *Fusicladium*, *Cladosporium*, *Exserohilum*, *Bipolaris*, *Cercospora*, *Alternaria*, and *Fusarium*, respectively.
3. Compile a retrieve table of the Anamorphic Fungi involved in this experiment.
4. Talk about the relationship between the Anamorphic Fungi and Ascomycota.

实验十一　无性型真菌重要植物病原物的识别与鉴定(Ⅱ)
——腔孢纲重要属形态观察

【概述】

腔孢纲(Coelomycetes)真菌的特征是分生孢子梗着生在分生孢子盘或分生孢子器内，分为黑盘孢目(Melanconiales)和球壳孢目(Sphaeropsidales)。将分生孢子果为分生孢子盘的真菌归属在黑盘孢目内，而为分生孢子器类型的真菌归属在球壳孢目内(图 2-10)。本纲中有不少真菌是重要的植物病原菌，在被害植物的发病组织表面上往往会形成黑色的小颗粒。

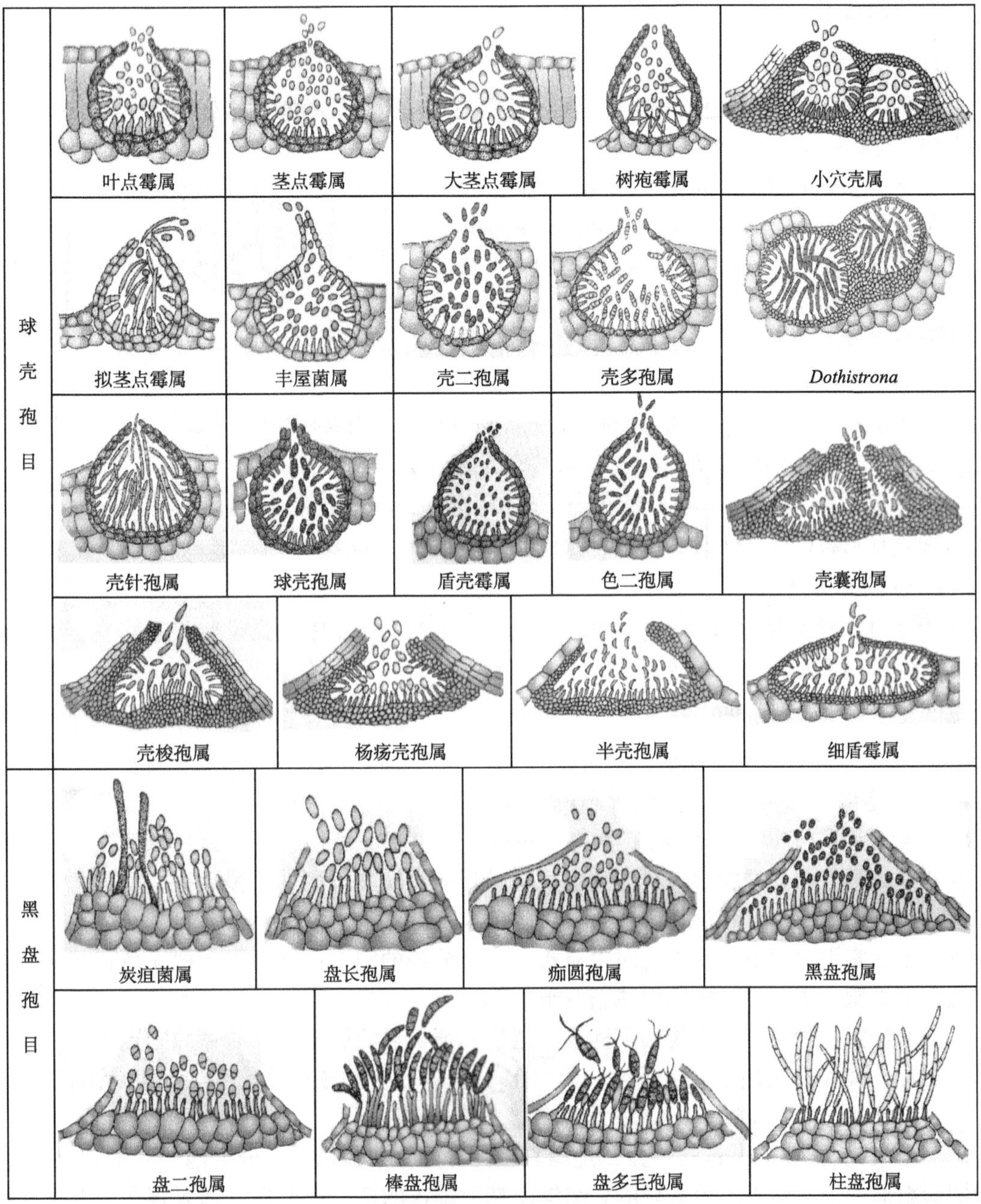

图 2-10 腔孢纲重要植物病原物的形态特征

（引自 Agrios，2005）

【实验目的】

1. 掌握腔孢纲真菌的形态特征和引起植物病害的症状特点。
2. 掌握腔孢纲重要病原物的鉴定特征。
3. 熟悉徒手切片制作技术。

【材料和器具】

1. 实验材料

准备下列植物病原物的培养物、永久显微玻片或引起的植物病害标本。

(1)黑盘孢目

①炭疽菌属(*Colletotrichum*):高粱炭疽病(*C. graminicola*)、正木炭疽病(*C. gloeosporioides*)、山茶炭疽病(*C. camelliae*)和棉花炭疽病(*C. gossypii*)。

②盘二孢属(*Marssonina*):花椒落叶病(*M. zanthoxyli*)、杨树黑斑病(*M. populicola*)、月季黑斑病(*M. rosae*)和苹果褐斑病(*M. mali*)。

③拟盘多毛孢属(*Pestalotiopsis*):松针赤枯病或枇杷灰斑病(*P. funerea*)。

(2)球壳孢目

①茎点霉属(*Phoma*):甜菜蛇眼病(*P. betae*)和葡萄黑腐病(*P. viticola*)。

②大茎点霉属(*Macrophoma*):苹果/梨轮纹病(*M. kawatsukai*)。

③叶点霉属(*Phyllosticta*):桂花赤枯病(*P. osmanthicola*)。

④壳针孢属(*Septoria*):菊花褐斑病(*S. chrysanthemella*)。

2. 仪器及用具

计算机、投影仪、显微镜、手持放大镜、刀片、载玻片、盖玻片、通草、镊子、剪刀、滴瓶、蒸馏水、吸水纸等。

【方法和步骤】

1. 观察所提供的由腔孢纲真菌引起的植物病害的症状，掌握其症状识别要点。

2. 腔孢纲重要病原物的显微形态观察与区分。

①炭疽菌属(*Colletotrichum*):常将由炭疽菌属引起的植物病害称为炭疽病，其症状的主要特征为具有明显的轮纹斑，后期在病斑处形成轮纹状排列的小黑点——病菌的分生孢子盘；当湿度大时，常在病斑上产生红褐色的黏性分泌物，为病菌的分生孢子团。分生孢子盘产生在寄主表皮下，盘内有时生有褐色、具分隔的刚毛；分生孢子梗无色至褐色；分生孢子单细胞，无色，长椭圆形或新月形。

选取任意一种炭疽病植物病害标本，选取病征明显的病斑，以通草为固定媒介，制作徒手切片，镜检，观察分生孢子盘、分生孢子梗和分生孢子的形态特点，以及刚毛的有无。边观察边绘制形态图。

②盘二孢属(*Marssonina*):盘二孢属真菌的分生孢子盘极小；分生孢子双细胞，无色，卵圆形至椭圆形，大小不等，分隔处缢缩。选取提供的任一种由盘二孢属真菌引起的植物病害标本，选取病征明显的病斑，制作徒手切片，镜检，观察分生孢子盘和分生孢子的形态特点。边观察边绘制形态图。

③拟盘多毛孢属(*Pestalotiopsis*):拟盘多毛孢属真菌的分生孢子由 5 个细胞构成，真隔膜，两端细胞无色，中间细胞橄榄褐色，顶生附属丝 2 根以上。以松针赤枯病或枇杷灰斑病的标本为材料，选取病征明显的病斑，制作徒手切片，镜检，观察分生孢子盘和分生孢子的形态特点。边观察边绘制形态图。

④茎点霉属(*Phoma*):该属真菌的分生孢子器球形，内部着生分生孢子梗，梗极短；

分生孢子单细胞，无色，较小，卵形至椭圆形。以甜菜蛇眼病标本为材料，选取病征明显的病斑，制作徒手切片，镜检，观察分生孢子器和分生孢子的形态特点。边观察边绘制形态图。

⑤大茎点霉属(*Macrophoma*)：该属真菌产生的分生孢子器和分生孢子形态与茎点霉属相似，但是分生孢子比茎点霉属的大，其直径一般均超过 15 μm。以苹果/梨轮纹病为材料，选取病征明显的发病组织，制作徒手切片，镜检，观察分生孢子器和分生孢子的形态特点。边观察边绘制形态图。

⑥叶点霉属(*Phyllosticta*)：该属真菌在形态上与茎点霉属相似，比较难以区分。传统上把寄生在茎上的该类病原真菌归属为茎点霉属，把寄生在叶片上的该类真菌归属为叶点霉属。取桂花赤枯病的病叶，选取病征明显的病斑，制作徒手切片，镜检，观察分生孢子器和分生孢子的形态特点。边观察边绘制形态图。

⑦壳针孢属(*Septoria*)：该属真菌的分生孢子器球形，分生孢子多细胞，无色，细长针形或线状，直或微弯。取菊花褐斑病的病叶，选取病征明显的病斑，制作徒手切片，镜检，观察分生孢子器和分生孢子的形态特点。边观察边绘制形态图。

【结果和讨论】

1. 总结供试植物病害标本的症状特点，掌握腔孢纲真菌引起植物病害的诊断要点。

2. 分别绘制炭疽菌属、盘二孢属和拟盘多毛孢属的分生孢子盘、分生孢子梗和分生孢子形态图。

3. 分别绘制茎点霉属、大茎点霉属、叶点霉属和壳针孢属的分生孢子器、分生孢子梗和分生孢子形态图。

4. 编制一检索表，区分腔孢纲重要病原物。

EXPERIMENT 11 Recognition and Identification of Important Plant Pathogens of Anamorphic Fungi(Ⅱ)

——Morphological Observation of Important Genera in Coelomycetes

【Introduction】

The characteristic of Coelomycetes is that conidiophores are produced in acervulus or pycnidium. Coelomycetes can be divided into Melanconiales and Sphaeropsidales. The fungi with conidiocarp as acervulus are categorized into Melanconiales, while the fungi with conidiocarp as pycnidium are categorized into Sphaeropsidales (Figure 2-10). Many fungi of Coelomycetes are important plant pathogens, which can usually form black little particles on diseased tissues of infected plants.

【Experimental Purpose】

1. Master the morphological characteristics of the fungi in Coelomycetes and the symptom characteristics caused by them.

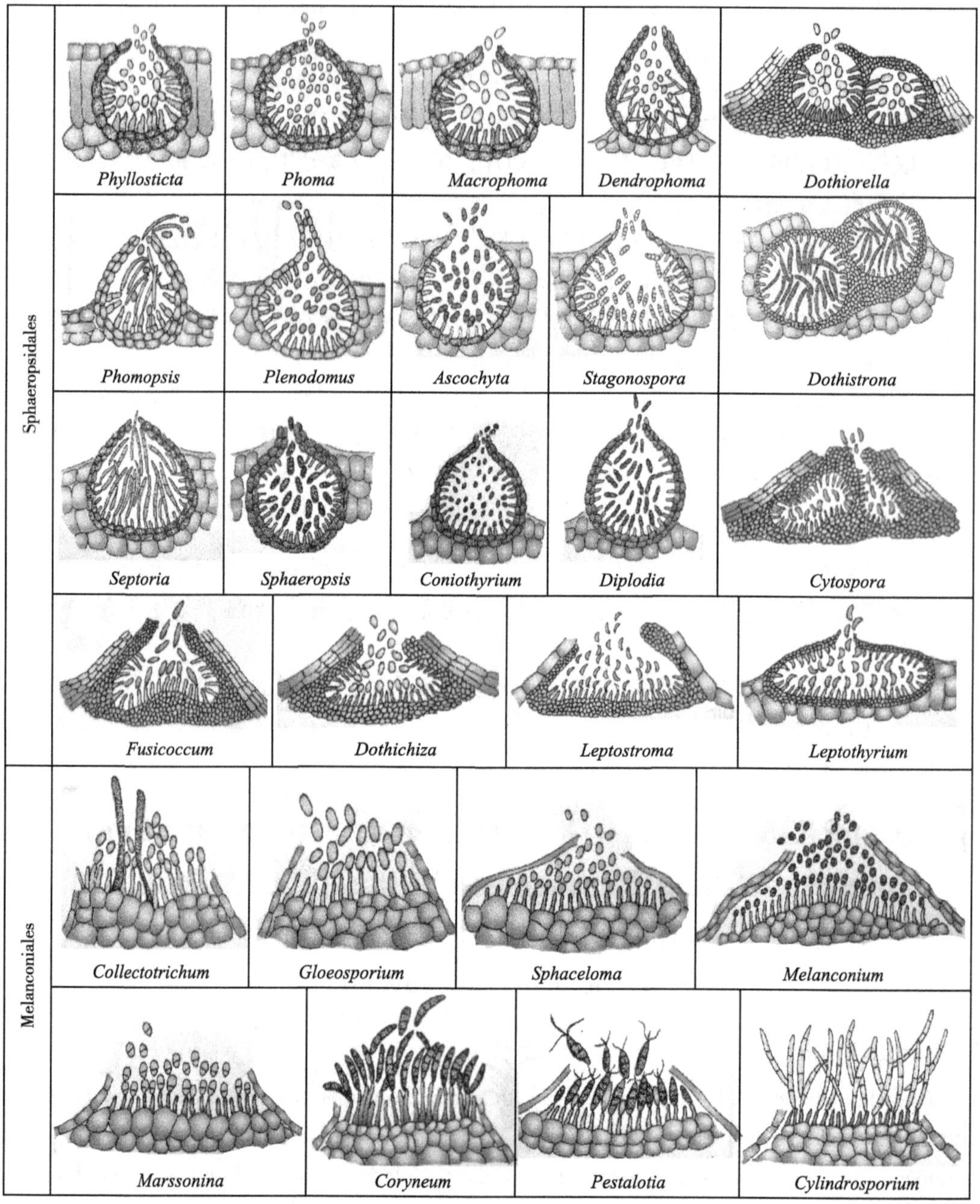

Figure 2-10 Morphologic Characteristics of Important Pathogens in Coelomycetes

(Cited from Agrios, 2005)

2. Master the identification characteristics of the important pathogens in Coelomycetes.
3. Be familiar with the technique of hand-making section.

【Materials and Apparatus】

1. Materials

Prepare the cultures in *vitro*, permanent slides or plant disease specimens caused by the following pathogens.

(1) Melanconiales

① *Colletotrichum*: Anthracnose of sorghum (*C. graminicola*), anthracnose of *Euonymus japonicus* (*C. gloeosporioides*), anthracnose of camellia (*C. camelliae*) and anthracnose of cotton (*C. gossypii*).

② *Marssonina*: Leaf cast of prickly ash (*M. zanthoxyli*), black spot of populus (*M. populicola*), black spot of Chinease rose (*M. rosae*), and brown spot of apple (*M. mali*).

③ *Pestalotiopsis*: Red blight of pine needles or gray spot of loquat (*P. funereal*).

(2) Sphaeropsidales

① *Phoma*: Clamp rot of sugar beet (*P. betae*) and black rot of grape (*P. viticola*).

② *Macrophoma*: Ring rot of apple/pear (*M. kawatsukai*).

③ *Phyllasticta*: Red blight of osmanthus (*P. osmanthicola*).

④ *Septoria*: Brown spot of chrysanthemum (*S. chrysanthemella*).

2. Instruments and Appliances

Computers, projectors, microscopes, handheld magnifying glasses, blades, slides, cover slides, ricepaperpbant piths, tweezers, scissors, dropping bottles, distilled water, blotting papers, etc.

【Methods and Procedures】

1. Observe the symptoms of all the provided plant disease specimens caused by the fungi of Coelomycetes, and master the key chracteristics for symptom recognition.

2. Observe and differentiate the microscopic morphology of the important pathogens in Coelomycetes.

① *Colletotrichum*: The plant diseases caused by *Colletotrichum* are usually named as anthracnose, which is mainly characterized by obvious ring spot and small black particles (i. e. the acervulus of pathogen) also arranged like a circle on the spot in the later stage of disease development. The reddish-brown sticky secretions are often produced on the spot under a high humidity, which are the red brown conidiospore mass. The acervulus is formed under the epidermis of host plant, in which brown septate seta are sometimes produced. Coniophores are colourless to brown. Conidiospores are unicellular, colourless and oblong or crescent in shape.

Take any plant disease specimen of anthrocnose, choose the spot with obvious sign, and then make handworked sections taking pith as the fixing medium to observe the morphological characteristics of acervulus, conidiophore and conidiospore, as well as the presence of seta or not under a microscope. Draw the morphological illustrations as you observe.

② *Marssonina*: The acervuli for the *Marssonia* fungi are extremely small, and conidiospores are bicellular, colorless, ovoid to elliptic in shape, different in size, and with constriction of septa. Take any plant disease specimen caused by *Marssonia*, and then choose the diseased spot with obvious sign to make handworked sections to observe the morphological characteristics of acervuli and conidiospores under a microscope. Draw the morphological illustrations as you observe.

③*Pestalotiopsis*: Each conidiospore for the *Pestalotiopsis* fungi is composed by five cells with true septa, and the cells on both terminals are colorless while the middle cells are olive brown. On the top of each conidiospore, more than two appendages are formed on the top. Take the specimen of red blight of pine needles or gray spot of loquat as material, and then choose the diseased spot with obvious sign to make fhandworked sections to observe the morphological characteristics of acervuli and conidiospores under a microscope. Draw the morphological illustrations as you observe.

④*Phoma*: The pycnidia for the *Phoma* fungi are spherical in shape, in which the extremely short conidiophores are formed. Conidiospores are unicellular, colorless, small, and oval to elliptic in shape. Take the specimen of clamp rot of sugar beet as material, and then choose the diseased spot with obvious sign to make handworked sections to observe the morphological characteristics of pycnidia and conidiospores under a microscope. Draw the morphological illustrations as you observe.

⑤*Macrophoma*: The morphology of pycnidia and conidiospore produced by *Macrophoma* is similar with that of *Phoma*. However, the size of conidiospore for *Macrophoma* is larger than that of *Phoma*, the diameter of which is usually more than 15 μm. Take the specimen of ring rot of apple or pear as material, and then choose the diseased tissues with obvious sign to make handworked sections to observe the morphological characteristics of pycnidia and conidiospores under a microscope. Draw the morphological illustrations as you observe.

⑥*Phyllosticta*: The genus of *Phyllosticta* is morphologically similar to the genus of *Phoma*, both of which are extremely difficult to distinguish. Traditionally, the fungi parasitizing in stem are categorized into *Phoma*, while the fungi parasitizing in leaf are usually identified as *Phyllosticta*. Take the diseased leaf of red blight of osmanthus as material, and then choose the spot with obvious sign to make handworked sections to observe the morphological characteristics of pycnidia and conidiospores under a microscope. Draw the morphological illustrations as you observe.

⑦*Septoria*: The pycnidia of *Septoria* are spherical in shape, and the conidiospores are multicellular, colorless, and elongated needle like or linear in shape, straight or slightly curved. Take the diseased leaf of brown spot of chrysanthemum, and then choose the spot with obvious sign to make handworked sections to observe the morphological characteristics of pycnidia and conidiospores under a microscope. Draw the morphological illustrations as you observe.

【Results and Discussion】

1. Summarize the symptom characteristics of all provided plant disease specimens, and master the key points for the diagnosis of plant disease caused by fungi in Coelomycetes.

2. Draw the morphological illustrations of acervuli, conidiophores and conidiospores respectively for the genera of *Colletotrichum*, *Marssonina* and *Pestalotiopsis*.

3. Draw the morphological illustrations of pycnidia, conidiophores and conidiospores respectively for the genera of *Phoma*, *Macrophoma*, *Phyllosticta* and *Septoria*.

4. Compile a retrieve table to distinguish the important pathogens in Coelomycetes.

实验十二 植物病原真菌的分离与培养

【概述】

植物的患病组织内存在大量的病原物，如果给予适宜的环境条件，除个别种类外，一般都能恢复其生长和繁殖。然而，患病组织部位的病原菌常常和其他微生物混杂在一起，采用组织分离培养法可将病原物从发病组织中分离出来，并通过多次纯化培养可将病原物与其他混杂微生物分开。病原物的分离和纯化是进行植物病原鉴定、分类和生物学特性研究的基础。

【实验目的】

1. 了解植物病原真菌分离及纯化的基本原理和步骤。
2. 掌握植物病原真菌的组织分离培养法操作技术。
3. 掌握单孢分离法的原理和操作技术。

【材料和器具】

1. 实验材料

①分离材料：花椒干腐病(*Fusarium zanthoxyli*)新发病枝干、花椒落叶病(*Marssonia zanthoxyli*)新发病叶片。

②培养基：马铃薯葡萄糖琼脂培养基(potato dextrose agar，PDA)；2%水琼脂培养基(water agar，WA，2 g 琼脂 +100 mL 水)。

③试剂：70%乙醇、95%乙醇、0.1%的升汞溶液、无菌水等。

2. 实验器具

无菌培养皿(直径 90 mm)、酒精灯、解剖刀、手术剪、镊子、试管、烧杯(10 mL、150 mL)、滤纸、记号笔、纱布、接种针、牙签、玻璃涂布器、研钵、移液器等。

【方法和步骤】

1. 分离前的准备工作

①工作环境的清洁和消毒：植物病原物的分离一般在无菌室或超净工作台中进行。无菌室和超净工作台清洁后用化学药物(84 消毒液或 5%石炭酸溶液)喷雾或紫外线照射(20~30 min)消毒。实验人员宜穿着灭菌后的工作服，戴上口罩，用肥皂洗手后用 70%乙醇擦手。

②分离用具的消毒：凡与分离材料接触的器具都必须进行消毒处理，可将这些用具浸于 95%乙醇中，使用前在火焰上灼烧进行灭菌。重复 2~3 次，再次使用时必须再次灭菌。

培养皿、试管等可采用干热灭菌法（160~170℃，60 min）灭菌，而培养基及蒸馏水则采用高压蒸汽灭菌法（121℃，20 ~30 min）灭菌。

③分离材料的选择：通常采用新发病的植物器官或组织为分离材料，可以减少腐生菌的污染。分离材料的选择是植物病原物成功分离的关键，因为在植物发病组织的坏死细胞中，除了病原物外，常常存在别的腐生微生物，所以一般选择发病植物组织部位的病健交界处为分离材料。

2. 组织分离法

①制备培养基平板：在超净工作台内，将经高压灭菌后尚未凝固的PDA培养基倒入培养皿内，每皿约15 mL，轻轻晃动培养皿，使培养基在培养皿内分布均匀。培养基冷却凝固后，即为PDA平板。

②取样：采集新鲜的花椒干腐病发病枝干，用自来水冲洗后，选择具有典型症状的病斑，先用灭菌手术刀剥去病斑树皮的角皮层，再用无菌解剖刀从病斑的病健交界处切取韧皮层组织小块（每边长5~7 mm），作为分离材料。

③分离材料的表面消毒：将植物组织小块放入小烧杯中，向小烧杯中倒入适量70%乙醇处理5~8 s，以除去组织表面的气泡，倒出乙醇；然后向小烧杯内倒入适量的0.1%升汞溶液进行表面消毒0.5~5.0 min，消毒结束后回收升汞溶液。

注意：采用升汞进行表面消毒的时间因分离材料而异，对于柔嫩的植物组织，处理时间可短些，反之则可长些。表面消毒的时间要严格控制，如果时间太长，升汞会渗透到组织内，杀死病原菌；而时间太短，则表面消毒不彻底，分离到的杂菌较多。因此，当分离一种新的植物病害的病原物时，可以设置不同的表面消毒时间，通过分析不同消毒时间对分离结果的影响，从而确定最佳表面消毒条件。根据经验，嫩叶的消毒时间一般为0.5~1.0 min；质厚的老叶为1.5~3.0 min；组织较厚的病树皮为3~5 min。

④漂洗：将经消毒的植物组织块转移至装有无菌水的无菌大烧杯中（约40 mL）漂洗30 s左右，倒出无菌水，再向烧杯中添加等体积的无菌水再次清洗，重复清洗3~4次。

⑤去除组织块表面水分：将植物组织块从最后一次漂洗后的无菌水中取出，放置在多层无菌滤纸上，以吸附组织表面的水分。

注意：滤纸放入玻璃培养皿中，用牛皮纸包装培养皿，经高压蒸汽灭菌，60℃烘干后使用。

⑥植物组织块摆放：将植物组织块转移至PDA平板上，轻轻按压，每个平板中放置3~4个植物组织块，呈三角形或正方形摆放（图2-11）。

⑦培养：用记号笔在培养皿底部注明植物病害名称、分离日期和学生姓名，将培养皿倒置（皿底在上），放置于恒温箱中，在25℃暗培养3~4 d。

⑧纯化：当观察到有微生物从植物组织块边缘长出时，选取典型的单菌落，用无菌接种针（或镊子）挑取带有菌丝的小块培养基，转移至新的PDA平板上培养，纯化培养2~3次后，可获得纯菌株。

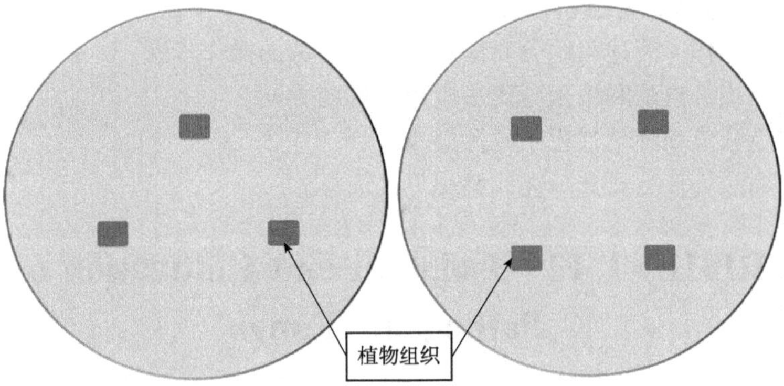

图 2-11　植物病原真菌组织分离法示意图

⑨保存：获得的纯菌种可保存在冰箱中(4℃，短期保存)，为后续的致病性验证提供接种菌种。

3. 单孢分离法

①制备培养基平板：在超净工作台内，将经高压灭菌后尚未凝固的 2% WA 培养基倒入培养皿内，每皿约 10 mL，轻轻晃动培养皿，使培养基在培养皿内分布均匀。培养基冷却凝固后即为 WA 平板。

②采样：采集新鲜的花椒落叶病的病叶，先用自来水冲洗，自然风干后，剪取密生分生孢子盘的病斑。

③消毒：表面消毒同组织分离法，消毒时间控制在 0. 5~1. 5 min。

④漂洗：同组织分离法。

⑤研磨：将漂洗后的植物组织块转移至无菌研钵中，加入 1~2 mL 无菌水，用研杵在病斑表面轻轻研磨促使内分生孢子释放，制备分生孢子悬浮液。

⑥涂布及培养：将所得分生孢子悬浮液稀释至 100 个/mL 左右(血球计数器计数)，取 20 μL 涂布于 2% WA 平板上，放置于恒温箱中，在 25℃暗培养 12 h。

⑦单孢挑取及培养：在光学显微镜下寻找培养皿中萌发的单个孢子，用记号笔标记。然后用无菌解剖针将标记的分生孢子连同琼脂块移入 PDA 平板上，放置于 25℃恒温箱中进行暗培养，1 个月后观察记录菌落大小和繁殖体产生情况，获得纯菌种。

⑧保存：获得的纯菌种可保存在冰箱中(4℃，短期保存)，为后续的致病性验证提供接种菌种。

4. 注意事项

①为了防止细菌污染，可以在培养基中添加适量的硫酸链霉素抑制细菌生长。

②所有分离操作均需要在无菌环境中(超净工作台)完成。

③在将植物组织块转移至培养基平板前，务必用无菌吸水纸或滤纸吸去组织表面水分，可以大大降低细菌污染的可能性。

【结果和讨论】

1. 汇报花椒干腐病和花椒落叶病的病原物分离结果并拍照打印，包括是否分离到病原物、是否存在污染等，并分析原因。

2. 根据实际操作，简述组织分离法和单孢分离法的操作步骤。

3. 试比较组织分离法和单孢分离法应用条件的异同。

4. 如何避免分离和纯化过程中的污染？

5. 组织分离法为什么要选择在病健交界处取样？

EXPERIMENT 12 Isolation and Cultivation of Plant Pathogenic Fungi

【Introduction】

There are massive individuals of pathogen in plant diseased tissues, which generally can be recovered to grow and reproduce under the appropriate environmental except for a few species. However, the pathogen in the diseased tissues usually exists combined with other microorganisms. The pathogen can be isolated from the infected plant tissues using the method of tissue isolation and culture, which is also can be purified by multiple purification cultivation to separate the pathogen from other miscellaneous microorganisms. Isolation and purification of pathogen is the foundation for the identification, classification and the researches on biological characteristics.

【Experimental Purpose】

1. Understand the basic principles and procedures of isolation and purification of plant pathogenic fungi.

2. Master the operation techniques of tissue isolation and culture of plant pathogenic fungi.

3. Master the principle and operation techniques of the method of single spore separation.

【Materials and Apparatus】

1. Materials

①Separation materials: fresh diseased stems of stem canker of *Zanthoxylum bumngeanum* (*Fusarium zanthoxyli*) and fresh diseased leaves of leaf cast of *Z. bungeanum* (*Marssonia zanthoxyli*).

②Medium: Potato dextrose agar(PDA) and 2% water agar medium(WA, 2 g agar+100 mL water).

③ Reagent: 70% alcohol, 95% alcohol, 0.1% mercury bichloride solution, sterile water, etc.

2. Instruments and Appliances

Sterile Petri dish (diameter 90 mm), alcohol lamp, scalpel, surgical scissors, tweezers, test tubes, beakers (10 mL and 150 mL), filter paper, marking pen, gauze, inoculating needle, toothpick, glass spreader, mortar, pipettors, etc.

【Method and Procedure】

1. Preparations for Separation

①Cleaning and disinfection of working environment: The isolation of plant pathogen is generally carried out in a asepsis room or an ultra-clean workbench. The asepsis room and the ultra-clean workbench should be cleaned, and then disinfected by spraying chemical agents(84 disinfectant or 5% carboniferous acid solution) or ultraviolet radiation(20-30 min). It's better to wear disinfected lab coat and mouth mask for the experimenters, wash hands using soap, and then wipe hands using 70% alcohol.

②Disinfection of separation tools: All tools in contact with the separation materials should be disinfected, which are immersed into 95% alcohol and burn on the flame to disinfect repeatedly 2-3 times before use. The tools should be disinfected again if they are used again. Petri dishes and test tubes can be sterilized by the method of hot-air sterilization(160-170℃, 60 min), while the medium and distilled water are sterilized by autoclaving(121℃, 20-30 min).

③Selection of separation materials: Commonly, the fresh diseased plant organs or tissues are taken as separation materials, which can reduce the pollution of saprophytes. The selection of separation material is the key to successfully isolate plant pathogen. There are other saprophytes existing in the necrotic cells besides pathogen, therefore, the junction of diseased and healthy sites of the infected plant tissues is usually taken as the separation material.

2. Tissue Separation Method

①Preparation of medium plate: In the ultra-clean workbench, pour the autoclaved PDA medium that has not yet solidified into the petri dishes for about 15 mL per dish, and gently shake the petri dishes to make the medium evenly distributed in the dishes. The cooled and solidified medium is also called PDA plate.

②Sampling: Collect the fresh diseased branches of stem canker of *Z. bungeanum*, wash them using running water, select the lesions with obvious symptom, peel off the cortex of bark of the lesion using a disinfected scalpel, and then cut some small pieces of phloem tissue(with each side 5-7 mm long) from the junction of diseased and healthy sites of lesions which are taken as separation materials.

③Surface disinfection for separation materials: Put the plant tissue pieces into a small beaker, and then pour a proper amount of 70% alcohol into the small beaker for 5-8 s treatment to remove the bubbles on the surface of the tissue, and then pour out the alcohol. An appropriate amount of 0.1% mercury bichloride solution is then added into the small beaker for surface disinfection for 0.5-5.0 min. The mercury bichloride solution can be recovered when the disinfection is finished.

Note: The treatment time for surface disinfection using mercury bichloride solution is dependent on the separation materials. For the tender plant tissues, the treatment time can be shorter, and vice versa. The treatment time for surface disinfection should be control strictly. If the treatment time is too long, mercury bichloride will penetrate into the tissues to kill the pathogens, while if it is too short, it will result in a halfway surface disinfection to isolate too many miscellaneous microorganisms. Therefore, when confronting a new plant disease to isolate its pathogen, different treatment time for surface disinfection can be designed. The optimal conditions for surface disinfection can be determined by analyzing the effects of different disinfection time on the isolation results. Building on experience, the disinfection time for young leaves is generally 0.5-1.0 min, for thick old leaves it is 1.5-3.0 min, while for the diseased thick bark, it is usually 3-5 min.

④ Rinse: Transfer the disinfected plant tissue pieces to a large sterile beaker containing sterile water (about 40 mL sterile water) for rinsing about 30 s, after that, pour out sterile water and add equal volume of sterile water to the beaker for washing 3-4 times repeatedly.

⑤Removement of water on the surface of tissue pieces: Take out the plant tissue pieces from the sterile water of the last time for rinsing, and place them on multiple layers of filter papers to remove the water adhering to the surface of plant tissues.

Note: Put the filter papers into the glass Petri dish, package the Petri dish using kraft paper, and then autoclave them, which are dried at 60℃ and then used.

⑥Placement of plant tissue pieces: Transfer the plant tissue pieces onto PDA plates and press them gently. Place 3-4 plant tissue pieces in each plate, showing the placement like triangle or square (Figure 2-11).

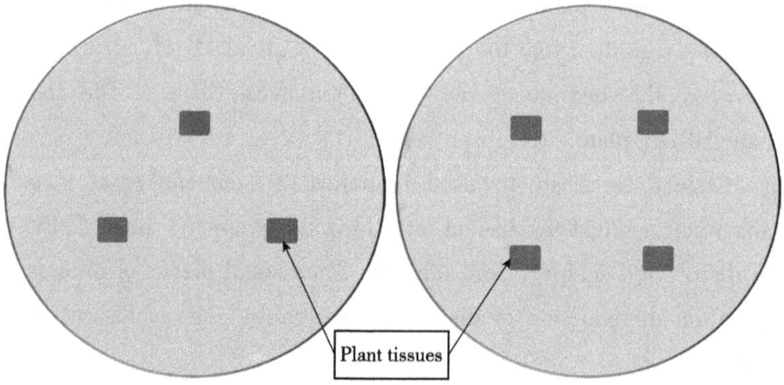

Figure 2-11 Schematic Diagram of Tissue Separation of Plant Pathogenic Fungi

⑦Cultivation: Use a marking pen to note plant disease name, isolation date and student name on the bottom of the Petri dish. Reverse the Petri dish over with the bottom of the dish on the upside, and place it in an incubator for a dark cultivation of 3-4 days under 25℃.

⑧Purification: When the microorganisms growing from the edges of plant tissue pieces can be seen, select a typical single colony, and then use a sterile inoculation needle (or tweezer) to pick up a small medium block with mycelia, which is then transferred onto a new PDA plate for cultivation. The pure isolates can be obtained by 2-3 times of purified cultivation.

⑨Preservation: The obtained pure isolates can be stored in a refrigerators (at 4℃ for a short-term storage), as the inoculation stains for the subsequent pathogenicity verification.

3. Single Spore Separation Method

①Preparation of medium plate: In the ultra-clean workbench, pour the autoclaved 2% WA medium that has not yet solidified into the petri dishes for about 10 mL per dish, and gently shake the petri dishes to make the medium evenly distributed in the dishes. The cooled and solidified medium is also called WA plate.

②Sampling: Collect the fresh diseased leaves of leaf cast of *Z. bungeanum*, wash them using running water, dry them naturally in air, and then take the spots with dense acervuli using a scissor.

③Disinfection: The surface disinfection is conducted as the same as the method of tissue isolation. The disinfection time is controlled within 0.5-1.5 min.

④Rinse: It is the same as the method of tissue isolation.

⑤Grinding: Transfer the rinsed plant tissue pieces to a sterile mortar, add 1-2 mL sterile water, and then grind gently on the surface of the spot with a pestle to release the conidiospores to prepare the spore suspension.

⑥Spreading and cultivation: Dilute the spore suspension to about 100 spores per milliliter (counting using a hemacytometer), take an aliquot of 20 μL to spread on a 2% WA plate, and then place it in an incubator for a dark cultivation of 12 h under 25℃.

⑦Picking up of single spore and cultivation: Find out the single spore in the state of germination using an optical microscope, mark it using a marking pen, and then transfer the marked conidiospore mixed with agar block to a PDA plate using a sterile dissecting needle, which is then placed under 25℃ for dark cultivation. One month later, record the size of colony and the reproductive bodies, and obtain pure isolate.

⑧Preservation: The obtained pure isolates can be stored in a refrigerators (at 4℃ for a short-term storage), as the inoculation stains for the subsequent pathogenicity verification.

4. Precautions

①To avoid bacterial contamination, an appropriate amount of streptomycin sulfate can be added to the medium to inhibit bacterial growth.

②All separation operations should be carried out in a sterile environment (ultra-clean workbench).

③Before placing plant tissue pieces on the medium plate, it is necessary to use disinfected absorbent papers or filter papers to absorb water on the surface of the tissue, which can greatly reduce the possibility of bacterial contamination.

【Results and Discussion】

1. Report the isolation results of the pathogens of *Z. bungeanum* stem canker and leaf cast, take photos and print them, including whether or not the pathogen is obtained and there is

contamination and then analyze the reasons.

2. Briefly describe the operation steps of the methods of tissue separation and single spore separation according to actual operations.

3. Discuss the similarities and differences of the application conditions between tissue separation method and single spore separation method.

4. How to avoid contamination in the separation and purification process?

5. Why should the junction of diseased and healthy sites be chosen for the method of tissue isolation?

实验十三　植物细菌病害的诊断及病原物的分离

【概述】

植物细菌病害的主要症状为坏死类型，有时也可见萎蔫、畸形等症状。在发病初期，受侵植物组织上常可见水渍状、半透明的局部性病斑，随着病程的发展，病斑颜色加深，变为褐色或黑褐色。叶部病斑常因受叶脉的限制而呈多角形。在输导组织中定植和扩展的细菌，常使叶脉和维管束坏死而变为黑褐色，在输导组织的切口断面稍加挤压可见溢脓现象。由细菌引起的畸形类植物病害，则罕见变色和组织坏死等症状。植物细菌性病害在高湿条件下，可在病斑表面观察到菌脓，常呈白色、黄色或粉红等各种颜色。溢脓现象是植物细菌病害诊断的重要依据。然而，在湿度不足时，许多细菌病害都很难出现溢脓，因此，需进行室内显微镜镜检，以查明发病组织中是否存在细菌，才能做出准确的诊断。

对于常见的植物细菌病害，可根据其症状特点对其进行初步的诊断。在此基础上，可对植物细菌病害的病原物进行分离，并通过分析病原物的形态学特征、生化特性和分子生物学特征，鉴定植物病原细菌，明确其分类地位。由于患病植物组织内细菌菌体量过多，所以，常采用稀释分离法进行植物病原细菌的分离。

【实验目的】

1. 认识植物细菌病害的症状类型及特点。
2. 掌握植物细菌病害的显微镜检查方法。
3. 掌握植物病原细菌的分离与培养技术。

【材料和器具】

1. 实验材料

①植物细菌病害标本：核桃细菌性黑斑病(*Xanthomonas campestris* pv. *juglandis*)新鲜发病叶片、猕猴桃细菌性溃疡病(*Pseudomonas syringae* pv. *actinidiae*)新鲜发病叶片或白菜软腐病(*Erwinia carotovora* subsp. *carotovora*)新鲜发病叶片。

②培养基：Luria-Bertani 琼脂(LBA)培养基。

③试剂：70%乙醇、95%乙醇、0.1%的升汞溶液和无菌水等。

2. 实验器具

显微镜、无菌培养皿、酒精灯、解剖刀、手术剪、尖头镊子、试管、烧杯(10 mL 和

150 mL）滤纸、记号笔、纱布、接种环、牙签、玻璃涂布器、研钵、移液器和枪头等。

【方法和步骤】

1. 植物细菌病害症状观察

观察供试植物细菌病害标本的症状特点，注意病斑上是否有菌脓。

2. 植物细菌病害的“喷菌”现象观察

大多数植物细菌病害在发病组织中均有大量细菌菌体，当对发病组织制作徒手切片进行显微镜观察时，可观察到大量的细菌菌体从发病组织中喷出，即“喷菌”现象，其为植物细菌病害所特有，是区分细菌或真菌、病毒病害较为简便的手段之一。维管束病害的喷菌量较多，可持续几分钟到十几分钟；薄壁组织病害的喷菌量较少，喷菌状态持续时间也较短。具体实验操作步骤如下：

①将供试病害标本的病叶在自来水下反复冲洗，以去除病叶表面的灰尘或腐生菌。

②用剪刀在病叶上剪取病斑，放置在干净的载玻片上，然后用刀片将病斑切割为边长为 3 mm 见方的正方形小块。

③用尖头镊子挑取 1～2 个组织块，放置于新的载玻片上，滴加 1 滴无菌水，盖上盖玻片，在光学显微镜下观察是否有“喷菌”现象。

3. 植物病原细菌的分离与培养

①分离材料的选择：经过镜检诊断的发病组织材料，即可作为分离材料。如为软腐病，则需在表面消毒的健康组织上重新接种，然后再在新发病组织上分离病菌。

②消毒：在发病植物组织病健交界处切取约 5 mm 见方的病组织小块，放入小烧杯内，加入一定量的 0.1%升汞溶液消毒处理 1 min，回收升汞消毒液。

③漂洗：用无菌水漂洗经升汞溶液消毒后的病组织，重复漂洗 3～4 次。

④研磨：将病组织块转移至无菌研钵中，添加 4～5 mL 无菌水，将病组织块研磨匀浆，静置 10～15 min 后，吸取一定量的上清液于无菌离心管中，即为细菌菌悬液，也可将其进一步稀释后使用。

⑤涂布：吸取 20～40 μL 细菌菌悬液于 LBA 平板上，用无菌玻璃涂布器将其均匀涂布在培养基平板上，也可用无菌接种环蘸取细菌菌悬液，采用划线法将菌液涂抹于 LBA 平板上，可呈“Z”字形涂布。

⑥培养：将 LBA 平板放置于恒温箱中，在 27℃暗培养 48～72 h 后，观察是否有细菌菌落长出，如菌落的颜色和形状均表现一致，说明分离成功。选择特征一致的优势菌落进行菌种保存。

⑦保存：从优势菌落中选取典型的单菌落，用无菌接种环将其转移至无菌水中进行稀释，制备细菌悬浮液，然后取 20 μL 细菌悬浮液置于 LBA 平板上，用无菌涂布器将其均匀地涂布在平板上，待单菌落长出后，挑取单菌落接种在新的 LBA 平板上，27℃暗培养 48 h 后，将其放置于 4℃冰箱内保存。

【结果和讨论】

1. 详细记录植物病原细菌的分离结果，包括是否分离到病原物、是否存在污染等，并分析原因。

2. 简述植物病原细菌的分离培养步骤及其相关的注意事项。
3. 根据柯赫氏法则，说明症状诊断及病菌显微镜检查的意义。
4. 分析植物病原真菌和细菌的分离方法的异同。

EXPERIMENT 13 Diagnosis and Pathogen Isolation of Plant Bacterial Diseases

【Introduction】

The main symptom of plant bacterial diseases is tissue necrosis, and sometimes the symptoms of wilt and malformation can be observed. At the initial stage of plant disease, it can be seen some watery and translucent local spots on the plant infected tissues. As the development of pathogenesis, the color of spots deepens to brown or dark brown. Leaf spots are usually polygonal in shape because of the limitation of leaf vein. The bacteria colonizing and spreading in the transduction tissues often lead to the necrosis of leaf veins and vascular bundles to become dark brown, and the phenomenon of pyorrhea can be observed when pressing on the incision section of transduction tissues. There are rarely the symptoms of discoloration and necrosis for the plant disease showing malformation caused by bacteria. The ooze is often observed on the surface of spots for plant bacterial diseases under the condition of high moisture, which is usually white, yellow or pink in color. The phenomenon of pyorrhea is an important basis for the diagnosis of plant bacterial diseases. However, pyorrhea is hardly observed for most plant bacterial diseases because of the insufficient humidity. Therefore, the observation under a microscope in the lab is necessary to check the existence of bacteria in the diseased tissues, and then an accurate diagnosis can be obtained.

Many common plant bacterial diseases can make a preliminary diagnosis according to the symptom characteristics, based on which the pathogen of plant bacterial diseases can be isolated. Furthermore, by analyzing the morphological, biochemical and biological characteristics of the pathogen, the bacterial pathogen can be identified and its taxonomic status also can be demonstrated. Commonly, the isolation of plant bacterial pathogen can be conducted using the method of dilution separation due to large quantity of bacteria individuals in the infected plant tissues.

【Experimental Purpose】

1. Understand the types and characteristics of symptoms of plant bacterial diseases.
2. Master the method of microscope observation of plant bacterial diseases.
3. Master the techniques of isolation and cultivation of plant pathogenic bacteria.

【Materials and Apparatus】

1. Materials

①Specimens of plant bacterial diseases: The fresh diseased leaves of walnut bacterial black spot(*Xanthomonas campestris* pv. *juglandis*), kiwifruit bacterial canker(*Pseudomonas syringae* pv. *actinidiae*) or Chinese cabbage soft rot(*Erwinia carotovora* subsp. *carotovora*).

②Medium: Luria-Bertani Agar(LBA) medium.

③Reagents: 70% alcohol, 95% alcohol, 0.1% mercury bichloride solution, sterile water, etc.

2. Instruments and Appliances

Microscopes, sterile Petri dishes, alcohol lamps, scalpels, surgical scissors, pointed forceps, test tubes, beakers(10 mL and 150 mL), filter papers, marking pens, gauzes, inoculating loops, toothpicks, glass spreaders, mortars, pipettes, pipette tips, etc.

【Methods and Procedures】

1. Symptom Observation of Plant Bacterial Diseases

Observe the symptom characteristics of all provided plant bacterial disease specimens, and note whether there is ooze on the spots or not.

2. Observation of "Bacteria Exudation" of Plant Bacterial Disease

There are massive bacterial thalli in infected tissues for most plant bacterial diseases. When the infected tissues are made as handworked sections to observe under a microscope, it can be observed that large quantity of bacterial thalli gush from infected tissues, i. e. the phenomenon of "bacteria exudation". The phenomenon of "bacteria exudation" is unique for plant bacterial diseases, which is one of the most convenient ways to differentiate plant bacterial disease from fungal and viral disease. It shows more amount of bacteria exudation for vascular bundle diseases, which can last several to more than 10 min. For parenchymatous diseases, there is a relatively less amount of bacteria exudation with a shorter duration. The specific experimental operation steps are as follows:

①The diseased leaves of the provided plant disease specimen are washed under running water repeatedly to remove the dust and saprophytes on the surface of diseased leaves.

②Use a scissor to cut the spots on the diseased leaves, put them on a clean glass slide, and then cut the spots as some square patches with the edge length as 3 mm using a blade.

③Pick up one to two patches using a pointed forceps to place them on a new glass slide, add one drop of sterile water, cover a cover glass and then observe under a microscope whether there is the phenomenon of "bacteria exudation".

3. Isolation and Cultivation of Plant Pathogenic Bacteria

①Selection of separation materials: The infected tissues diagnosed by microscopic examination can be used as the separation materials. For soft rot, re-inoculation is necessary on the healthy tissues that has been disinfected on the surface, and then isolate the pathogen from the new infected tissues.

②Disinfection: Cut some small square pieces of the diseased tissue with the edge length as 5 mm from the junction between healthy and diseased tissues, put them in a small beaker, and then add a certain amount of 0. 1% mercuric chloride solution for disinfection for 1 min. The used mercuric chloride solution can be recycled.

③Rinse: Wash the diseased tissues disinfected by mercuric chloride solution with sterile water repeatedly for 3-4 times.

4. Grinding

Transfer the diseased tissue species to a sterile mortar, add 4-5 mL of sterile water, and then grind the diseased tissue species to a homogenate. After a standing time of 10-15 min, draw a certain amount of supernatant to a sterile centrifuge tube, i. e. the bacterial suspension, which can also be further diluted for use.

①Spreading: Take an aliquot of 20-40 μL bacterial suspension onto the LBA plate, and then spread it on the medium plate using a sterile glass spreader uniformly. It is also feasible to use an inoculating loop to dip into the bacterial suspension, and then employ the streak method to spread it on the LBA plate as the type of "Z".

②Cultivation: All LBA plates are placed in an incubator, which are cultured at 27℃ in dark for 48-72 h, and then observe whether there are bacterial colonies growing on the plate. If the color and shape of colonies are the same, it indicates the isolation is successful. The dominant colonies with the same characteristics are selected for bacteria preserving.

③Preservation: Select a typical single colony from the dominant colonies, and then use a sterile inoculating loop to transfer it into sterile water to dilute for the preparation of bacteria suspension. Take an aliquot of 20 μL of bacteria suspension, transfer it onto the LBA plate, and then spread it on the plate uniformly using a sterile glass spreader. When there are single colonies growing on the plate, pick up the single colony and inoculate it on a new LBA plate, which is then cultivated in dark at 27℃ for 48 h and then placed in a refrigerator for preservation at 4℃.

【Results and Discussion】

1. Record the isolation results of plant pathogenic bacteria in detail, including whether or not the pathogen is contamination and there is contamination, and analyze the reasons.

2. Briefly describe the steps for the isolation and cultivation of plant pathogenic bacteria and its related precautions.

3. Explain the significance of symptom diagnosis and pathogen microscopic examination according to Koch's postulate.

4. Analyze the the similarities and differences of the isolation methods between plant pathogenic fungi and bacteria.

实验十四　植物病原物的人工接种及病程观察

【概述】

从发病植物上分离到的微生物是否为病原物，需要采用科赫法则对其致病性进行验证才能得以确认，即将该分离物接种在寄主植物上，观察其是否能诱发植物产生与原来发病植物相同的症状，并从接种发病的植物上再次分离到该分离物。

接种是指人为地将病原物与寄主植物感病部位接触，为病原物创造条件使其成功侵入植物组织中，并诱导寄主发病。利用人工接种技术验证病原物的致病性，是开展病原物鉴定、病害发展规律调查、植物抗病性鉴定和病原物致病性测定等相关研究的前提。因此，人工接种技术是植物病理学研究工作中的一项基本技能，也是科赫法则中的关键步骤。

植物病害的人工接种方式，可根据病原物的传染方式和侵染途径进行设计。植物病害的种类众多，其传染方式和侵染途径各异，因此，接种方法也不尽相同。有时因研究目的不同，需要采用特殊的接种方法。植物侵染性病害的发生是由病原物、寄主植物和环境条件3个因素共同决定的，植物病害的人工接种系统能否成功建立与这3个因素直接相关。因此，在设计人工接种方法时，必须考虑这3个因素的作用，如接种体的活力和致病性、寄主的生长期及感病程度、环境的湿度及温度等。

【实验目的】

1. 掌握植物病原物致病性验证的基本原理。
2. 学习和掌握植物病原物的常用接种方法和操作技能。
3. 了解不同接种方法对植物病害发生发展过程的影响。

【实验材料和器具】

1. 实验材料

①病原菌：花椒干腐病（*Fusarium zanthoxyli*）、落叶松—杨栅锈（*Melampsora larici-populina*，MLP）、苹果褐斑病（*Marssonia mali*）、核桃细菌性黑斑病（*Xanthomonas campestris* pv. *juglandis*）和花椒根腐病（*Fusarium solani*）。

②供试植物：花椒和杨树盆栽苗、苹果和核桃新鲜叶片。

③培养基：PDA培养基、PCDA培养基（10 g土豆，10 g胡萝卜，2 g葡萄糖，0.7 g琼脂，100 mL蒸馏水）、LBA培养基。

2. 实验器具

酒精灯、显微镜、血球计数器、无菌水、解剖针、镊子、脱脂棉、保鲜膜、喷雾瓶、塑料花盆、保湿装置、记号笔、标签、锥形瓶、剪刀、营养土、毛笔等。

【方法和步骤】

1. 针刺接种法

由伤口侵入植物组织病原物可采用针刺接种法将其接种到寄主植物上。通常用无菌解剖针在寄主植物接种部位制造微伤口，然后将病原菌接种在伤口上。可采用该接种方法接种花椒干腐病病原菌，具体步骤如下：

①将保存于−80℃的花椒干腐病病原菌菌种取出，接种于PDA平板上，放置于25℃的培养箱内，暗培养10~14 d。

②用圆形无菌打孔器在培养了14 d的花椒干腐病病原菌菌落边缘上制备直径为5 mm

的圆形菌饼。

③首先用湿的无菌棉擦拭花椒的健康枝干，再用酒精棉擦拭花椒枝干上设定的接种位点，所有接种位点均设定为直径约 5 mm 的圆形位点，与菌饼大小相当。然后用无菌解剖针在每个接种位点上个针刺 7~9 微伤口。

④用无菌牙签挑一个菌饼接种于接种位点上，菌丝面贴合伤口。切记不能移动菌饼的位置。然后用浸泡在无菌水中的脱脂棉片包裹接种位点，再用保鲜膜包裹每个接种位点以进行保湿处理；以在微伤口上接种 PDA 培养基作为对照。

⑤所有接种的花椒盆栽苗放置于植物温室中培养，培养温度为 25℃，相对湿度为 75%，光周期为 12 h/d，光照强度为 2 000 lx。在接种后 14 d，揭开保鲜膜和脱脂棉，观察记录花椒干腐病的发病情况。

2. 涂抹接种法

涂抹接种法是指将病原菌孢子悬浮液直接涂抹在寄主植物表面，通常用来接种可直接侵入或经自然孔口和伤口侵入植物体内的病原物。本实验以落叶松—杨栅锈菌(MLP)为例，主要步骤如下：

①将保存于-80℃的 MLP 夏孢子粉放入培养皿中，称重并记录。

②对 MLP 夏孢子粉进行吸湿处理 6~8 h ，以使其活化，然后用无菌水配制成孢子悬浮液，浓度为 1~2 mg/mL。

③用清水洗去杨树叶面上的尘土等脏物，然后用喷雾瓶在杨树叶背面喷雾，以制造一层水膜。

④用毛笔蘸取 MLP 夏孢子悬浮液，均匀涂抹于杨树叶背面；以在叶片上涂抹无菌水为对照。

⑤将接种的杨树植株放置于保湿桶内避光保湿 24 h 后，将植株取出，放置于温度为 25℃的温室中培养。

⑥分别于接种后 24 h、48 h、72 h、120 h 和 168 h 观察记录杨树叶锈病的发病情况。

3. 悬滴接种法

悬滴接种法是将病菌的悬浮液滴在寄主植物表面。经气流和雨水传播的病原物都可以用此接种方法。以苹果褐斑病病原菌的接种为例，主要步骤如下：

①将保存于-80℃的苹果褐斑病病原菌菌种取出，接种于 PCDA 平板上，放置于 25℃的培养箱内暗培养 1 个月。

②刮取苹果褐斑病病原菌的黑褐色菌落，转移至研钵中，添加 2 mL 无菌水，将其研磨成菌悬液。然后用毛笔将菌悬液涂抹于经表面消毒的苹果叶片背面，25℃保湿暗培养，待有分生孢子从叶片表面溢出时，用毛笔蘸取无菌水反复刷洗叶片上的分生孢子，制备成分生孢子悬浮液，并用血球计数器和显微观察将孢子悬浮液的浓度调整至 1×10^6个/mL 左右。

③采集健康的苹果叶片，先用自来水冲洗叶片，再用 6% 的次氯酸钠进行表面消毒，最后用无菌水重复漂洗 3~5 次。将叶片放置于垫有无菌滤纸的灭菌圆形的玻璃培养皿中(直径 15 cm，高度 2. 5 cm)，叶柄末端用潮湿的无菌棉包裹。

④用移液枪吸取 20 μL 分生孢子悬浮液，以悬滴的方式接种于苹果叶背面；以在叶面

上接种无菌水为对照。

⑤盖上培养皿盖，用封口膜封口，置于培养箱中在25℃条件下进行光照培养(光照12 h/d)，5~7 d后观察记录苹果褐斑病的发病情况。

4. 剪叶接种法

该方法可用于接种经水孔或伤口侵入的植物病原菌，以核桃细菌性黑斑病病原菌为例，主要步骤如下：

①取出保存于冰箱中的核桃细菌性黑斑病的病菌，采用划线培养法将其接种在LBA平板上活化培养24~48 h。

②挑取LBA平板上的单菌落，接种于LB液体培养基中，在27℃振荡培养24 h后，用血球计数器在显微镜下对细菌菌悬液计数，并将其浓度调整为1×10^7 CFU/mL左右。

③采集健康的核桃叶片，先用自来水冲洗叶片表面，再用6%的次氯酸钠进行表面消毒，最后用无菌水重复漂洗3~5次。将叶片放置于垫有无菌滤纸的经过灭菌的玻璃培养皿中(直径15 cm，高度2. 5 cm)，叶柄末端用蘸有无菌水的脱脂棉包裹。

④用无菌剪刀蘸取细菌悬浮液，然后用剪刀剪核桃叶片，每个伤口长7~8 mm；以蘸取LB液体培养基后的剪刀剪切的叶片为对照。

⑤盖上培养皿盖，用封口膜封口，放置于培养箱中在25℃进行光照培养(光照12 h/d)，5~7 d后观察记录核桃细菌性黑斑病的发病情况。

5. 土壤接种法

植物根部病害的病原菌常采用该方法接种。以花椒根腐病病原菌为例，主要步骤如下：

①将保存于−80℃的花椒根腐病病原菌菌种取出，接种于PDA平板上，放置培养箱内，在25℃条件下暗培养5~7 d。

②向培养有花椒根腐病病原菌菌落的培养皿中添加5~10 mL无菌水，用无菌棉签轻轻擦拭菌落表面，制备孢子悬浮液，在显微镜下用血球计数器对其计数，并将孢子悬浮液的浓度调整为1×10^6个/mL左右。

③以2年生健康花椒盆栽苗为接种植物，将适量的花椒根腐病病原菌的孢子悬浮液灌入花椒根部的土壤中，以接种相同体积无菌水的花椒苗为对照。

④所有接种的花椒盆栽苗放置于植物温室中培养，培养温度为25℃，相对湿度为75%，光周期为12 h/d，光照强度为2 000 lx。每隔3 d观察一次花椒盆栽苗的生长情况，包括叶片是否变色、是否脱落等，于接种后21 d将花椒盆栽苗拔出土壤，观察记录花椒根腐病的发病情况。

【结果和讨论】

1. 记录接种实验的全过程，详细描述和分析接种实验结果。
2. 针对不同的病原物，如何选择合适的接种方法？
3. 分析影响病原物侵入寄主植物组织的影响因素。
4. 结合自己开展的具体实验，谈谈植物病原物成功接种的关键环节及注意事项。

EXPERIMENT 14 Artificial Inoculation of Plant Pathogen and Pathogenesis Observation

【Introduction】

It is need to be confirmed whether the isolated microorganism from diseased plant is the pathogen or not by applying Koch's postulate to verify its pathogenicity, that is to inoculate the isolate on the host plant, observe whether it can cause the same symptom as the original diseased plant, and obtain the isolate again from the inoculated diseased plant.

Inoculation refers to make the pathogen contact with the susceptible site of host plant artificially, which can create the conditions for the successful penetration of pathogen into plant tissues and cause disease of host. It is a premise that employing artificial inoculation to verify the pathogenicity of pathogen for the researches on the identification of pathogen, the investigation of disease development regularity, the identification of plant resistance and the detection of pathogen pathogenicity. Therefore, artificial inoculation is a basic technique for the researches on phytopathology, which is also the essential steps for Koch's postulate.

The ways of artificial inoculation of plant diseases can be designed according to the infectious types and penetration pathways of pathogens. The numerous types of plant diseases accompanied with varied infectious types and penetration pathways result in different inoculation methods. Sometimes, the special inoculation method is necessary because of different research aims. The occurrence of plant infectious disease is depends on the three factors, including pathogen, host plant and environmental conditions. Whether the artificial inoculation system of plant disease can be established successfully depends on the above three factors directly. Hence, the functions of the above three factors should be considered when designing the method of artificial inoculation, such as the vitality and pathogenicity of inoculum, the growth stage and susceptibility of host, and the environmental humidity and temperature, etc.

【Experimental Purpose】

1. Master the basic principle of pathogenicity verification of plant pathogens.
2. Learn and master the common inoculation methods and operation skills of plant pathogens.
3. Understand the effects of different inoculation methods on the occurrence and development of plant disease.

【Materials and Apparatus】

1. Materials

①Pathogens: The pathogen of prickly ash stem canker(*Fusarium zanthoxyli*), the pathogen of larch-poplar rust (*Melampsora larici-populina*, MLP), the pathogen of apple brown spot (*Marssonia mali*), the pathogen of walnut bacterial black spot (*Xanthomonas campestris* pv. *juglandis*), and the pathogen of prickly ash root rot(*Fusarium solani*).

②Provided plants: The potted seedlings of prickly ash and poplar, and the fresh leaves of apple and walnut.

③Media: PDA medium, PCDA medium (10 g potato, 10 g carrot, 2 g glucose, 0.7 g agar, 100 mL distilled water), LBA medium.

2. Instruments and Appliances

Alcohol lamps, microscopes, hemacytometers, sterile water, dissecting needles, tweezers, absorbable cotton, preservative film, spray bottles, plastic pots, moisturizing devices, marking pens, tags, Erlenmeyer flasks, scissors, nutrient soils, brushs, etc.

【Methods and Procedures】

1. Acupuncture Inoculation

The pathogen that penetrates into plant tissues via wound can be inoculated onto host plant by the method of acupuncture inoculation. Generally, a sterile dissecting needle can be used to make microwound on the inoculation site of host plant, and then the pathogen is inoculated onto the microwound. The pathogen of prickly ash stem canker can be inoculated by this method with the specific steps as follows:

①Take out the fungus *F. zanthoxyli* stored at −80℃, inoculate it on a PDA medium plate, and then place it in an incubator at 25℃ for dark cultivation for 10-14 d.

②A round plug containing *F. zanthoxyli* mycelia with a diameter of 5 mm was prepared from a 14-day-old colony of *F. zanthoxyli* using a round hole puncher.

③The healthy stems of *Z. bungeanum* were cleaned with wet sterilecotton firstly, and then the designed inoculation sites are wiped again by alcohol wipes. All inoculation sites are designed as round sites with the diameter of 5 mm, which is the same as mycelial plug in size. Next, each inoculation site is punctured by a sterile dissecting needle to create 7-9 tiny stab wounds.

④A mycelial plug is picked up by a sterile toothpick and inoculated on an inoculation site with the mycelial side attached to the microwound. Do remember not to move the location of mycelial plug. Next, each inoculation site is wrapped in absorbent cotton soaked in sterile water and sealed tightly with plastic film to maintain humidity. The microwound sites inoculated with only sterile PDA plugs are considered as negative controls (CK).

⑤All inoculated plants are cultured in a greenhouse under a temperature of 25℃, a relative humidity of 75%, and a photoperiod of 12 h/d with a light intensity of 2 000 lx. The observation and record of stem canker on *Z. bungeanum* stems is carried out at 14-day postinoculation after the plastic film and cotton are removed.

2. Smear Inoculation

The method of smear inoculation refers to apply the spore suspension of pathogen directly on the surface of host plant, which is usually used to inoculate those plant pathogens that can infect directly or via natural orifices and wounds. Taking the pathogen *Melampsora larici-populina* (MLP) as an example in this experiment, the main steps are as follows:

①Transfer the urediospores of MLP stored at −80℃ to a Petri dish, weigh it and record.

②The urediospores of MLP are subjected to hygroscopic treatment for 6-8 h to activate them, which are then made as spore suspension using sterile water with the concentration as 1-2 mg/mL.

③Remove the dust and dirt on the leaves of poplar with clean water, and then spray the back of the leaves with a spray bottle to create a layer of water film.

④Use a brush to dip into MLP urediospores suspension and smear the spore suspension on the back side of poplar leaves. The leaves smeared by sterile water are taken as the control.

⑤The inoculated poplar seedlings are placed in a moisturizing bucket to protect against light and moisturize for 24 h, after which take out the seedlings from the bucket and place them in a plant greenhouse for cultivation at 25℃.

⑥The occurrence of poplar leaf rust is observed and recorded at 24 h, 48 h, 72 h, 120 h and 168 h after inoculation.

3. Hanging Drop Inoculation

Hanging drop inoculation refers to add the suspension of pathogen dropwise on the surface of plant. The pathogens that are transmitted by air and rain can be inoculated by this method. Take the pathogen of apple brown spot as an example, the main steps are as follows:

①Take out the fungus *Marssonia mali* stored at −80℃, inoculate it on a PCDA medium plate, and then place it in an incubator at 25℃ for dark cultivation for one month.

②Scrape the black-brown colonies of *M. mali*, transfer them to a mortar, add 2 mL sterile water, and grind them to make the fungal suspension. Next, use a brush to dip into the suspension of *M. mali* and smear it on the back side of disinfected poplar leaves. The treated leaves are cultured at 25℃ in dark. When there are conidiospores secreted from the leaf surface, use a brush to dip into sterile water and wash out the conidiospores on the leaves repeatedly to make conidiospore suspension. The concentration of conidiospore suspension is adjusted as 1×10^6 spores /mL using a hemacytometer under the observation of microscope.

③Collect healthy apple leaves, wash them using running water, disinfect them with 6% sodium hypochlorite, and wash them with sterile water repeatedly for 3-5 times. Place the leaves in a sterile Petri dish (diameter 15 cm, height 2.5 cm) padded with sterile filter papers, and the terminals of petioles are wrapped with wet sterile cotton.

④An aliquot of 20 μL spore suspension is taken with a pipette and inoculated on the back side of apple leaves in the way of hanging drops. The leaves inoculated by sterile water are taken as control.

⑤The petri dishes are covered with lips, sealed with a sealing film, and placed in a incubator at 25℃ for light cultivation (light for 12 h/d). The occurrence of apple brown spot is observed and recorded after 5-7 d.

4. Leaf-cutting Inoculation

This method is suitable for the inoculation of pathogens that infect plants through hydathode or wounds. Taking the pathogen of walnut bacterial black spot as an example, the main steps are as follows:

①Take out the pathogen of walnut bacterial black spot stored at a refrigerator, and inoculate it on the LBA plate using the method of streak plating for the activating cultivation for 24-48 h.

②Pick up a single colony on the LBA plate, and inoculate it in LB liquid medium for a shaking cultivation at 27℃ for 24 h. Following, use a hemacytometer to calculate the concentration of the bacterial suspension under a microscope, and then adjust it to the concentration of 1×10^7 CFU/mL.

③Collect the healthy leaves of walnut, wash them using running water firstly, disinfect them with 6% sodium hypochlorite, and finally wash them with sterile water repeatedly for 3-5 times. Place the leaves in a sterile round glass Petri dish (diameter 15 cm, height 2.5 cm) padded with sterile filter papers, and the terminals of petioles are wrapped with wet sterile cotton.

④Dip a sterile scissor into the bacterial suspension, and then cut the walnut leaves with the scissors. Each wound is about 7-8 mm in length. The leaves cut by a scissors that is dipped in LB liquid medium are taken as the control.

⑤The petri dishes are covered, sealed with a sealing film, and placed in a incubator at 25℃ for light cultivation (light for 12 h/d). The occurrence of walnut bacterial black spot is observed and recorded after 5-7 d.

5. Soil Inoculation

The pathogens of plant root diseases are usually inoculated using this method. Taking the pathogen of prickly ash root rot as an example, the main steps are as follows:

①Take out the fungus *Fusarium solani* stored at −80℃, inoculate it on a PDA medium plate, and then place it in an incubator at 25℃ for dark cultivation for 5-7 d.

②Add 5-10 mL sterile water into the Petri dish containing the colony of *F. solani*, and then use an autoclaved cotton bud to wipe the surface of colony to make spore suspension. Next, use a hemacytometer to calculate the concentration of spore suspension under a microscope, and adjust it to the concentration of approximately 1×10^6 spores/mL.

③The two-year-old healthy prickly ash potted seedlings are taken as the inoculation plants. A certain amount of spore suspension of *F. solani* is injected into the soil around the root of prickly ash. The potted seedlings of prickly ash inoculated with the same volume of sterile water are taken as control.

④All inoculated plants are cultured in a greenhouse under a temperature of 25℃, a relative humidity of 75%, and a photoperiod of 12 h/d with a light intensity of 2 000 lx. The growth situations of prickly as potted seedlings are observed once at intervals of three days, including whether or not the leaves change color and fall off, ect. The observation and record of prickly ash root rot is carried out on 21-day postinoculation after the plastic film and cotton are removed. The occurrence of prickly ash root rot is observed and recorded on 21-day postinoculation after prickly ash potted seedlings are pulled out of the soil.

【Results and Discussion】

1. Record the total process of the inoculation experiment, and describe and analyze the results of the inoculation experiment in detail.

2. How to choose an appropriate inoculation method for different pathogens?

3. Analyze the factors influencing the invasion of pathogen into host tissues.

4. Talk about the key steps and precautions for the successful inoculation of pathogen based on the particular experiment carried out by yourself.

实验十五　植物病原真菌的 ITS 测序与鉴定

【概述】

真菌的传统分类鉴定以形态学为基础，然而，真菌种类多，形态学特征又极为复杂，形态学鉴定技术的掌握需要很强的专业背景，导致了基于形态学对真菌进行分类鉴定难度较大。近些年，随着分子生物学技术的快速发展，生物系统分类基础发生了重大变化，对微生物的分类鉴定已不再局限于表型特征，而是进入了表型特征与分子特征相结合的阶段。从分子水平上研究生物的分子特征，为微生物分类鉴定提供了简便、准确的技术和方法。

真菌的核糖体内部转录间隔区(internal transcribed spacer, ITS)位于 rDNA 片段上，由 ITS1、5.8S rDNA 和 ITS2 三个区域构成，不被转录翻译为蛋白质，分别与 18S rDNA 小亚基单元(SSU)和 28S rDNA 大亚基单元(LSU)相连，长度一般为 500~800 bp，常为多拷贝(图 2-12)。

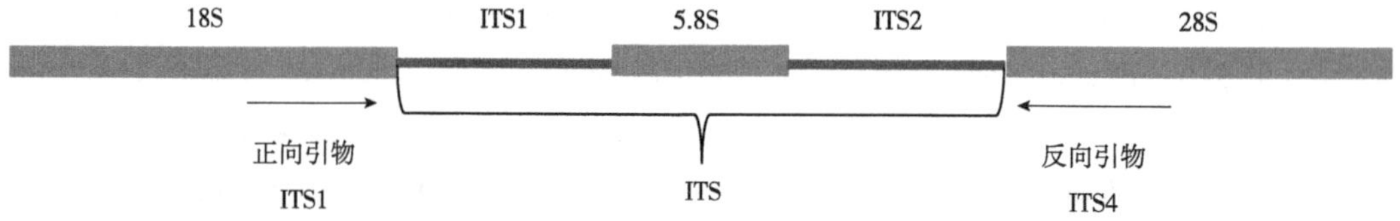

图 2-12　真菌的 ITS 序列及引物示意图

ITS 序列进化速率相对较快，种内高度保守，属间差异明显，引物通用性强，扩增效率高，这些特点使 ITS 成为真菌首选的 DNA“条形码”，广泛应用于真菌物种鉴定及系统进化分析。

【实验目的】

1. 了解利用 ITS 序列分析鉴定植物病原真菌的原理。

2. 掌握 PCR 扩增真菌 ITS 序列的原理与基本操作技术。

3. 掌握真菌系统发育树构建的技能和相关软件的使用方法。

【材料和器具】

1. 实验材料

①病原真菌：花椒根腐病(*Fusarium solani*)。

②实验试剂：十六烷基三甲基溴化铵(CTAB)、苯酚、氯仿、异丙醇、PCR缓冲液、2×ES Taq MasterMix、引物ITS1和ITS4、DNA Ladder 2000、琼脂糖、TBE电泳缓冲液、核酸染料Goldview、TE缓冲液、异戊醇、无水乙醇、ddH_2O(双蒸水)、PDA培养基等。

2. 实验器具

培养箱、超净工作台、电子天平、恒温水浴锅、漩涡振荡器、微波炉、PCR仪、移液器、离心管、PCR管、培养皿、电泳仪、电泳槽、凝胶成像系统、高速离心机等。

【方法和步骤】

1. 病原菌DNA的提取

病原菌DNA的提取可采用CTAB法，具体操作过程如下：

①以在PDA平板上生长了7~10 d的花椒根腐病培养物为材料，进行DNA提取。用无菌小勺将病原菌的菌丝体从PDA平板上刮下来，放置在研钵中，加入少量灭菌石英砂(或液氮)研磨成匀浆(或粉状)。

②向研钵加入0.5 mL经65℃水浴预热的CTAB溶液，研磨3~5 min。将研磨后的菌丝匀浆转移至一无菌离心管A中，向离心管A添加0.5 mL预热的CTAB，用漩涡振荡器混匀后，65℃恒温水浴1 h，每10 min颠倒摇匀一次。

③向离心管A加入1 mL苯酚：氯仿：异戊醇(25：24：1)DNA提取液，混匀，12 000 r/min室温离心10 min，取上清液于一新的离心管B中。

④向离心管B中加入等体积的预冷异丙醇，充分混匀，在-20℃冰箱中放置2.5 h(至少1 h)。

⑤将离心管B从冰箱中取出，12 000 r/min室温离心15 min，弃上清液，向离心管B加入适量(约1 mL)的70%乙醇(乙醇应在-20℃下提前预冷)，12 000 r/min室温离心10 min，弃上清液。

⑥向离心管B加入适量(约1 mL)的95%乙醇(乙醇应在-20℃提前预冷)，12 000 r/min室温离心10 min，弃上清液，将离心管B放置于超净工作台内，利用通风系统使其挥干残留的乙醇。

⑦向离心管B加入TE缓冲液(或ddH_2O，0.5 mL左右)溶解DNA，置于-20℃冰箱，备用。

2. PCR扩增与检测

① PCR扩增引物：ITS1(5′-TCCGTAGGTGAACCTGCGG-3′)和ITS4(5′-TCCTCCGCTTATTGATATGC-3′)，由生物公司合成。

②PCR扩增体系：PCR扩增总体积为30 μL，包括15 μL 2×ES Taq MasterMix、1 μL ITS1引物(10 μmol/mL)、1 μL ITS4引物(10 μmol/mL)、0.5 μL DNA模板和12.5 μL ddH_2O。

③PCR扩增程序：使用PCR仪进行PCR扩增，其扩增程序为94℃预变性5 min，然后依次按94℃变性30 s、56℃退火45 s、72℃延伸45 s的程序循环30次，最后72℃延伸10 min。

④PCR扩增产物的检测：PCR扩增产物用1%琼脂糖凝胶电泳后，使用凝胶成像仪对

ITS 的扩增结果进行可视化检测。根据 DNA Marker 条带的位置，判断 ITS 序列是否被成功扩增，扩增成功的 ITS 条带大小在 500~800 bp。

3. ITS 测序

将 ITS 的 PCR 扩增产物交测序公司完成测序。

4. ITS 序列分析与系统发育树的构建

①将测得的 ITS 序列在 NCBI 数据库中进行 BLAST 分析，以确定目标病原真菌的分类地位。

②将目标病原真菌的 ITS 序列提交至 GenBank，获得登录号。

③用 Clustal X 软件进行多重序列比对分析。

④利用建树软件如 Mega 7.0 构建系统发育树，对目标病原真菌的系统发育关系进行分析。

【结果和讨论】

1. 记录 ITS 的 PCR 扩增结果，并打印其凝胶电泳成像图。
2. 打印 ITS 的测序结果，分析其序列特征，鉴定目标病原真菌。
3. 将基于 ITS 序列构建的系统发育树打印出来，并对目标病原真菌的系统发育关系进行分析。
4. 谈谈利用 ITS 序列鉴定真菌的优点和缺点。
5. 谈谈可用于植物病原真菌的鉴定的分子“条形码”。

EXPERIMENT 15 ITS Sequencing and Identification of Plant Pathogenic Fungi

【Introduction】

The traditional classification and identification of fungi is based on morphology, however, it needs an extreme professional background to master the morphological identification technique due to the numerous species of fungi and their very complex morphological characteristics. Therefore, it is extremely difficult to classify and identify fungi based on morphology. In recent years, with the rapid development of the techniques and methods of molecular biology, the taxonomic base of organisms has been dramatically changed. The classification and identification of microorganisms has not been confined to phenotypic traits, while it enters into a period combining phenotypic traits with molecular features. Studying the characteristics of biological molecular on the molecular level can provide convenient and accurate techniques and methods for the classification and identification of microorganisms.

The internal transcribed spacer (ITS) of fungus locates on rDNA, consisting of the three regions of ITS1, 5.8S rDNA and ITS2, which can not be transcribed and translated into proteins. The region of ITS is linked to the 18S rDNA small subunit (SSU) and 28S rDNA large subunit (LSU), respectively, which is usually 500 to 800 bp in length and multiple copies

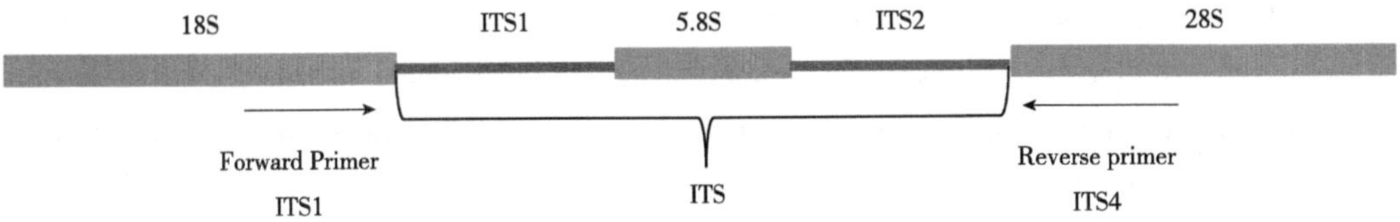

Figure 2-12　Sequence and Primers of ITS of Fungi

(Figure 2-12).

The characteristics of ITS sequences, such as relatively fast evolution, high conservation within the species, significant difference between genera and high application efficiency, make ITS become the preferred DNA "barcode" for fungi and has been widely applied for the identification of fungal species and phylogenetic analysis.

【Experimental Purpose】

1. Understand the principle of identification of plant pathogenic fungi by ITS sequence analysis.

2. Master the principle and basic operation techniques of PCR amplification of fungal ITS sequence.

3. Master the skills for the construction of fungal phylogenetic tree and the application methods of related softwares.

【Materials and Apparatus】

1. Materials

①Pathogenic fungus: The pathogen of prickly ash root rot(*Fusarium solani*).

②Reagents: Cetyl trimethyl ammonium bromide (CTAB), phenol, chloroform, isopropanol, PCR buffer, 2×ES Taq MasterMix, primers of ITS1 and ITS4, DNA Ladder 2000, agarose, TBE electrophoretic buffer, nucleic acid dye Goldview, TE buffer, isoamyl alcohol, anhydrous ethanol, ddH_2O, PDA medium, etc.

2. Instruments and Appliances

Incubators, clean benchs, electronic balances, thermostat water baths, vortex oscillators, microwave ovens, PCR instruments, pipettes, centrifugal tubes, PCR tubes, Petri dishes, electrophoresis apparatus, electrophoresis tanks, gel imaging systems, high-speed centrifuges, etc.

【Methods and Procedures】

1. DNA Extraction of Pathogen

The extraction of DNA from pathogen can be carried out by CTAB method. The specific operation steps are as follows:

①DNA extraction is carried out taking the culture of *F. solani* cultured on PDA plate for 7-10 days as material. The mycelia of pathogen are scraped off PDA plate using a sterile spoon and placed in a mortar. A small amount of autoclaved quartz sand (or liquid nitrogen) is added into the mortar to grind the mycelial as homogenate (or powder).

②Add an aliquot of 0.5 mL CTAB solution preheated at 65℃ in a water bath to the mortar and grind the mycelia for 3-5 min. The ground mycelial homogenate is transferred to a sterile centrifuge tube A, and an aliquot of 0.5 mL preheated CTAB is added to centrifuge tube A, which is then mixed with a vortex oscillator and placed in a thermostat water bath at 65℃ for 1 h. The tube A is reversed to mix once at intervals of 10 min.

③Add an aliquot of 1 mL of DNA extraction solution, i.e. phenol : chloroform : isoamyl alcohol(25 : 24 : 1), to tube A, and mix them. Tube A is centrifuged at 12 000 r/min at room temperature for 10 min, and the supernatant is transferred to a new centrifuge tube B.

④Equal volume of precooled isopropyl alcohol is added to tube B, thoroughly mixed and then stored in a refrigerator at −20℃ for 2.5 h(at least 1 h).

⑤Tube B is taken out from the refrigerator, and centrifuged at 12 000 r/min at room temperature for 15 min. The supernatant is discarded, and a suitable amount of(about 1 mL) of 70% alcohol (the alcohol should be precooled at −20℃) is added to tube B, which is then centrifuged again at 12 000 r/min at room temperature for 10 min. The supernatant is discarded.

⑥A proper amount(about 1 mL) of 95% alcohol(alcohol should be precooled in advance at −20℃) is added to tube B, which is then centrifuged at 12 000 r/min at room temperature for 10 min. The supernatant is discarded, and tube B is placed in a clean bench to volatilize the residual alcohol using the ventilating system.

⑦Add TE buffer solution(or ddH_2O, about 0.5 mL) to tube B to dissolve DNA, and place it in a refrigerator at −20℃ for later use.

2. PCR Amplification and Detection

①Primers for PCR amplification: ITS1 (5′-TCCGTAGGTGAACCTGCGG-3′) and ITS4 (5′-TCCTCCGCTTATTGATATGC-3′), which can be synthesized by a bio company.

②PCR amplification system: PCR amplification is conducted in 30 μL of mixture containing 15 μL 2 × ES Taq Mastermix, 1 μL ITS1 primer(10 μmol/mL), 1 μL ITS4 primer(10 μmol/mL), 0.5 μL DNA template and 12.5 μL ddH_2O.

③PCR amplification procedure: PCR amplification is performed in a PCR thermal cycler with an initial denaturing step at 94℃ for 5 min, followed by 30 amplification cycles of 30 s denaturation at 94℃, 45 s primers annealing at 56℃, 45 s extension at 72℃, and then a final elongation step of 10 min at 72℃.

④Detection of PCR products: PCR products are subjected to electrophoresis using a 1% agarose gel, after that, visual confirmation of the ITS region is performed by a gel imaging system. It can be estimated whether or not the ITS sequences are amplified successfully according to the positions of DNA marker. The length of ITS sequences amplified successfully locates between 500-800 bp.

3. ITS Sequencing

Sequence analysis of the PCR products of ITS is carried out by a sequencing company.

4. ITS Sequence Analysis and Phylogenetic Tree Construction

①ITS sequence is analyzedin NCBI database by using the program of BLAST, which can confirm the taxonomic status of target pathogenic fungus.

② ITS sequence of target pathogenic fungus is submitted to GenBank to obtain its GenBank ID.

③Multiple sequences alignment is conducted by Clustal X software.

④ Construct the phylogenetic tree using software such as Mega 7.0, and analyze the phylogenetic relationship of target pathogenic fungus.

【Results and Discussion】

1. Record the result of PCR amplification for ITS, and print its image acquired by the imaging capture device.

2. Print the results of ITS sequencing and analyze its characteristics to identify the target pathogenic fungus.

3. Print the phylogenetic tree based on ITS sequences and analyze the phylogenetic relationship of target pathogenic fungus.

4. Discuss the advantages and disadvantages using ITS sequences to identify fungi.

5. Talk about the other molecular "barcode" applied to identify plant pathogenic fungi.

实验十六 植物病原细菌的 16S rDNA 测序与鉴定

【概述】

细菌的传统鉴定主要以其形态特征和生化性状等表型为依据，烦琐费时，且有时鉴定结果不准确。近年来，随着核酸技术的飞速发展，核酸序列分析已经广泛应用于细菌鉴定、种系发生及分类，常用的方法如 G+C 含量分析、DNA 杂交、rDNA 指纹、质粒图谱、16S rDNA 序列分析等，其鉴定的准确性、灵敏性和快捷性均得到极大的提高。

16S rDNA 是编码原核生物核糖体小亚基 rRNA(16S rRNA)的 DNA 序列，存在于所有细菌基因组中。16S rDNA 大小适中，约 1.5 kb，信息量大，具有保守性和普遍性等特点，包含 10 个保守区和 9 个可变区，既能体现不同细菌属之间的差异，又便于序列分析。

16S rDNA 中可变区序列因细菌不同而异，恒定区基本保守，所以可以利用恒定区序列设计引物，通过 PCR 将细菌的 16S rDNA 片段扩增出来，利用可变区序列的差异对不同属、种的细菌进行分类鉴定。16S rDNA 序列分析技术的基本原理是从待鉴定的细菌中扩增出 16S rDNA 片段，通过测序获得其序列信息，再将其在 NCBI 数据库中进行 BLAST 分析，构建系统发育树，分析目标菌株与其他微生物之间在遗传进化过程中的亲缘关系，从而达到对其分类鉴定的目的。目前，16S rDNA 测序技术已被广发应用于细菌的分类鉴定。

【实验目的】

1. 了解利用 16S rDNA 序列分析来鉴定植物病原细菌的原理。

2. 掌握利用 16S rDNA 序列分析进行细菌鉴定的操作技能。

3. 掌握细菌系统发育树构建的技能和相关软件的使用方法。

【材料和器具】

1. 实验材料

①病原细菌：核桃细菌性黑斑病(*Xanthomonas campestris* pv. *juglandis*)。

②实验试剂：10% SDS 溶液、蛋白酶 K、氯化钠、CTAB/NaCl 缓冲液、苯酚、异戊醇、氯仿、异丙醇、TE 缓冲液、PCR 缓冲液、2×ES Taq MasterMix、DNA Ladder 2000、无水乙醇、通用引物 27F 和 1492R、琼脂糖、TBE 缓冲液、核酸染料、ddH_2O、LB 培养基等。

2. 实验器具

振荡培养箱、超净工作台、电子天平、恒温水浴锅、微波炉、PCR 仪、移液器、离心管、PCR 管、接种环、培养皿、电泳仪、电泳槽、凝胶成像仪、高速离心机等。

【方法和步骤】

1. 细菌基因组 DNA 的提取

细菌基因组 DNA 的提取采用 CATB/NaCl 法，具体操作过程如下：

①挑取核桃细菌性黑斑病病原菌单菌落接种于 5 mL LB 液体培养基中，于 37℃振荡培养过夜，作为病菌的种子培养液。

②取 1 mL 种子培养液加入含有 10 mL LB 培养基的离心管 A 中，在 37℃振荡培养 24 h，5 000 r/min 离心 10 min，弃上清液。

③向离心管 A 加入 10 mL TE 缓冲液重悬菌体，5 000 r/min 离心 10 min，弃上清液。

④向离心管 A 再次加入 10 mL TE 缓冲液重悬菌体，充分混合，备用。

⑤取 3. 5 mL 细菌悬浮液于 10 mL 离心管 B，添加 10%的 SDS 溶液 184 μL，混匀，再添加 37 μL 10 mg/mL 的蛋白酶 K，混匀，在 37℃孵育 1 h。

⑥向离心管 B 加入 740 μL 5 mol/L NaCl 和 512 μL CTAB/NaCl 缓冲液，充分混匀后 65℃孵育 10 min。

⑦向离心管 B 加入等体积的氯仿：异戊醇(24：1)，混匀后 10 000 r/min 离心 5 min，收集上清液于新的离心管 C 中。

⑧向离心管 C 中加入等体积的苯酚：氯仿：异戊醇(25：24：1)，混匀后 10 000 r/min 离心 5 min，收集上清液于新的离心管 D 中。

⑨向离心管 D 中加入 0. 6 倍体积的异丙醇，充分混匀，10 000 r/min 离心 5 min，弃上清液，收集 DNA 沉淀物，并用 70%乙醇离心洗涤 DNA 沉淀物。

⑩向离心管 D 添加 1 mL TE 缓冲液溶解 DNA，4℃保存，备用。

2. PCR 扩增及检测

①PCR 扩增引物：27F(5′-AGAGTTTGATCCTGGCTCAG-3′)和 1492R(5′-TACGGCTACCTTGTTACGACTT-3′)，由生物公司合成。

②PCR 扩增体系：PCR 扩增总体积为 30 μL，包括 15 μL 2×ES Taq MasterMix、1 μL 27F 引物(10 μmol/mL)、1 μL 1492R 引物(10 μmol/mL)、0. 5 μL DNA 模板和 12. 5 μL ddH_2O。

③PCR 扩增程序：94℃预变性 5 min，然后依次按 94℃变性 30 s、56℃退火 45 s、72℃延伸 1 min 的程序循环 30 次，最后 72℃延伸 10 min。

④PCR 扩增产物的检测：PCR 扩增产物用 1%琼脂糖凝胶电泳后，使用凝胶成像仪对 16S rDNA 的扩增结果进行可视化检测。根据 DNA Marker 条带的位置，判断 16S rDNA 序列是否被成功扩增，扩增成功的 16S rDNA 条带大小约为 1.5 kb。

3. 16S rDNA 测序

将 16S rDNA 的 PCR 扩增产物交测序公司完成测序。

4. 16S rDNA 序列分析与系统发育树的构建

①将测得的 16S rDNA 序列在 NCBI 数据库中进行 BLAST 分析，以确定目标病原细菌的分类地位。

②将目标病原细菌的 16S rDNA 序列提交至 GenBank，获得登录号。

③用 Clustal X 软件进行多重序列比对分析。

④利用建树软件如 Mega 7.0 构建系统发育树，对目标病原细菌的系统发育关系进行分析。

【结果和讨论】

1. 记录 16S rDNA 的 PCR 扩增结果，并将其凝胶电泳成像图打印出来。

2. 将 16S rDNA 的测序结果打印出来，并分析序列特征，鉴定目标病原细菌。

3. 将基于 16S rDNA 序列构建的系统发育树打印出来，并对目标病原细菌的系统发育关系进行分析。

4. 利用 16S rDNA 序列分析技术获得的鉴定结果与原来结果是否一致？若不一致，如何确定其准确的分类地位？

5. 谈谈利用 16S rDNA 序列鉴定细菌的原理及优势。

EXPERIMENT 16　Sequencing of 16S rDNA and Identification of Plant Pathogenic Bacteria

【Introduction】

The traditional identification of bacteria is mainly based on their morphological characteristics and biochemical features, which is time consuming, laborious and sometimes inconclusive. With the rapid development of nucleic acid technique, sequence analysis of nucleic acid has been widely applied for bacterial identification, phylogeny and classification. The common methods include G+C content analysis, DNA hybridization, rDNA fingerprint, plasmid map, 16S rDNA sequence analysis, and so on, which improve the accuracy, sensitivity, and rapidity greatly for bacterial identification.

16S rDNA is the DNA sequence encoding the small subunit rRNA (16S rRNA) of the prokaryotic ribosome, which is present in the genomes of all bacteria. The length of 16S rDNA is a moderate, about 1.5 kb, which contains abundant information. The sequence of 16S rDNA is characterized by conservation and universality, including ten conserved regions and nine variable regions, which not only can reflect the differences between different bacterial genera, but also is very convenient for sequence analysis.

The sequences of variable regions of 16S rDNA vary with different bacteria, while the sequences of constant regions are usually conserved. Universal primers can be designed according to the sequence of constant region, which can be used to amplify the 16S rDNA by PCR. Different genera and species of bacteria can be classified and identified based on the sequence differences of variable regions. The basic principle of 16S rDNA sequence analysis technique is to amplify the fragment of 16S rDNA from the target bacterium, sequence it to obtain its sequence information, analyze it in NCBI database by BLAST, and construct phylogenetic tree to analyze the phylogenetic relationship between the target bacterium and other microorganisms in the process of genetic evolution, which can realize the goal of bacterial classification and identification. So far, the technique of 16S rDNA sequencing has been widely applied for the classification and identification of bacteria.

【Experimental Purpose】

1. Understand the principle of the application of 16S rDNA sequence analysis to identify plant pathogenic bacteria.

2. Master the technique of the application of 16S rDNA sequence analysis for bacterial identification.

3. Master the skills for the construction of bacterial phylogenetic tree and the application methods of related softwares.

【Materials and Apparatus】

1. Materials

① Pathogenic bacterium: the pathogen of walnut bacterial black spot (*Xanthomonas campestris* pv. *juglandis*).

② Reagents: 10% SDS solution, proteinase K, sodium chloride, the buffer of CTAB/NaCl, phenol, isoamyl alcohol, chloroform, isopropanol, TE buffer, PCR buffer, 2×ES Taq MasterMix, DNA ladder 2000, absolute ethanol, universal primers of 27F and 1492R, agarose, TBE buffer, nucleic acid dye, ddH_2O, LB medium, etc.

2. Instruments and Appliances

Shaking incubators, clean benchs, electronic balances, thermostat water baths, microwave ovens, PCR instruments, pipettes, centrifugal tubes, PCR tubes, inoculating loops, Petri dishes, electrophoresis apparatus, electrophoresis tanks, gel imaging systems, high-speed centrifuges, etc.

【Methods and Procedures】

1. Extraction of Bacterial Genomic DNA

Genomic DNA of bacterium was extracted by the method of CATB/NaCl, and the specific operation procedures are as follows:

①A single colony of *X. campestris* pv. *juglandis* is picked up and inoculated in 5 mL LB liquid medium, which is then shaken overnight at 37℃ as a seed culture.

②An aliquot of 1 mL of seed culture is added into a centrifuge tube A containing 10 mL of LB medium, which is then shaken for a cultivation of 24 h at 37℃. Next, tube A is centrifuged at 5 000 r/min for 10 min, and the supernatant is discarded.

③Add 10 mL TE buffer solution to tube A to resuspend the bacteria, centrifuge it at 5 000 r/min for 10 min, and discard the supernatant.

④Add 10 mL TE buffer solution to tube A again to resuspend the bacteria, mix them thoroughly for later use.

⑤Take an aliquot of 3. 5 mL bacterial suspension into a new centrifuge tube B, add 184 μL 10% SDS solution, mix them, and then add 37 μL 10 mg/mL proteinase K, which is mixed again and incubated at 37℃ for 1 h.

⑥Add 740 μL 5 mol/L NaCl solution and 512 μL CTAB/NaCl buffer to tube B, and mix them thoroughly. Tube B is then incubated at 65℃ for 10 min.

⑦An equal volume of chloroform : isoamyl alcohol (24 : 1) was added to tube B, after mixing, it is centrifugated at 10 000 r/min for 5 min. The supernatant is collected in a new centrifuge tube C.

⑧An equal volume of phenol : chloroform : isoamyl alcohol(25 : 24 : 1) is added to tube C, after mixing, it is centrifugated at 10 000 r/min for 5 min. The supernatant is collected in a new centrifuge tube D.

⑨A certain amount of isopropyl alcohol is added to tube D, which is then mixed thoroughly and centrifuged at 10 000 r/min for 5 min. The supernatant is discarded and the DNA precipitate in tube D is collected which is then washed with 70% ethanol. The DNA precipitate is collected again after centrifugation.

⑩An aliquot of 1 mL TE buffer is added to tube D to dissolve DNA, which is then stored at 4℃ for later use.

2. PCR amplification and Detection of 16S rDNA

①Primers for PCR amplification: 27F(5′-AGAGTTTGATCCTGGCTCAG-3′) and 1492R(5′-TACGGCTACCTTGTTACGACTT-3′), which can be synthesized by a biocompany.

②PCR amplification system: PCR amplification is conducted in 30 μL of mixture containing 15 μL 2 × ES Taq Mastermix, 1 μL 27F primer(10 μmol/mL), 1 μL 1492R primer(10 μmol/mL), 0. 5 μL DNA template and 12. 5 μL ddH_2O.

③PCR amplification procedure: PCR amplification is performed in a PCR thermal cycler with an initial denaturing step at 94℃ for 5 min, followed by 30 amplification cycles of 30 s denaturation at 94℃, 45 s primers annealing at 56℃, 1 min extension at 72℃, and then a final elongation step of 10 min at 72℃.

④Detection of PCR products: PCR products are subjected to electrophoresis using a 1%

agarose gel, after which, visual confirmation of 16S rDNA region is performed by a gel imaging system. It can be estimated whether or not 16S rDNA sequences are amplified successfully according to the positions of DNA marker. The length of 16S rDNA sequences amplified successfully is about 1. 5 kb.

3. 16S rDNA Sequencing

Sequence analysis of the PCR products of 16S rDNA is carried out by a sequencing company.

4. 16S rDNA Sequence Analysis and Phylogenetic Tree construction

①16S rDNA sequence is analyzed in NCBI database by using the program of BLAST, which can confirm the taxonomic status of the target pathogenic bacterium.

②16S rDNA sequence of the target pathogenic bacterium is submitted to GenBank to obtain its GenBank ID.

③Multiple sequences alignment is conducted by Clustal X software.

④Construct the phylogenetic tree using software such as Mega 7. 0, and analyze the phylogenetic relationship of the target pathogenic bacterium.

【Results and Discussion】

1. Record the result of PCR amplification for 16S rDNA, and print its image acquired by the imaging capture device.

2. Print the results of 16S rDNA sequencing and analyze its characteristics to identify the target pathogenic bacterium.

3. Print the phylogenetic tree based on 16S rDNA sequences and analyze the phylogenetic relationship of the target pathogenic bacterium.

4. Is the identification using 16S rDNA sequence analysis consistent with the original results? If not, how to ensure its accurate taxonomic status?

5. Talk about the principles and advantages using 16S rDNA sequence to identify bacteria.

实验十七 植物病毒病的症状观察及病毒内含体检查

【概述】

植物病毒病的主要诊断依据是其症状特征，其症状主要有变色、坏死和畸形。植物病毒病在寄主组织外表无病征，但是有些植物遭受病毒侵害后，在受侵组织内部会形成一些结晶状或不定形的非晶体状的结构——内含体。不同属的植物病毒往往产生不同形状的内含体，通常有不定形、结晶状、风轮状等。晶体状内含体一般是由病毒粒体整齐排列堆叠而成，而不定形内含体则是由病毒粒体和寄主细胞成分混合组成的。存在于细胞核内的称为核内含体，存在于细胞质的称为细胞质内含体。内含体多存在于症状明显的植物组织中，大型内含体可在光学显微镜下观察，而较小的内含体则需在电子显微镜下才能看清。内含体的形态差异可作为植物病毒病诊断或病毒鉴定的一个重要参考依据，如马铃薯 Y 病毒属(*Potyvirus*)的病毒形成风轮状内含体，而烟草花叶病毒属(*Tobamovirus*)的病毒则形成晶板状内含体。

【实验目的】

1. 认识植物病毒病害的症状特性和主要症状类型。

2. 掌握病毒内含体的本质及撕片法制片观察技术。

【材料和器具】

1. 实验材料

植物病毒病害标本：烟草花叶病毒病(TMV)、蚕豆花叶病毒病(BBMV)、黄瓜花叶病毒病(CMV)、苹果花叶病毒病(AMV)、小麦土传花叶病毒病(SbWMV)、烟草环斑病毒病(TRSV)、番茄条纹病毒病(TSV)、马铃薯卷叶病毒病(PVY)、番茄蕨叶病毒病(TFLV)和大麦黄矮病(BYDV)。

2. 实验试剂

①鲁戈尔碘液：称取1.0 g碘和2.0 g碘化钾，放入研钵中研磨，研磨时缓慢加入适量的无菌水直至碘和碘化钾溶解，然后将溶解液转移至容量瓶中，添加无菌水将其定容至300 mL，混匀后，转移至棕色试剂瓶，黑暗保存，备用。

②台盼蓝染液：称取0.5 g台盼蓝，溶解于100 mL加热的0.9 mol/L NaOH溶液作为原液，使用时稀释2 000~5 000倍。

③焰红染液：又名荧光桃红，称取1 g焰红溶解于100 mL水中。

3. 实验器具

显微镜、载玻片、盖玻片、尖头镊子、刀片、浮载剂等。

【方法和步骤】

1. 植物病毒病害的症状观察

①变色：观察烟草花叶病毒病、蚕豆花叶病毒病、黄瓜花叶病毒病、苹果花叶病毒病和小麦土传花叶病的标本，记录其症状特点，如叶片的变色特征、叶片的厚薄及是否有缩叶。注意观察这几种病毒病害引起的变色是否存在差异，是否有病征。

②坏死：观察烟草环斑病毒病和番茄条纹病毒病的标本，记录这两种坏死病斑的特点。注意观察是否有病征。

③畸形：观察马铃薯卷叶病毒病、番茄蕨叶病毒病和大麦黄矮病的标本，记录其症状特点。注意发病组织与健康组织的形态差异，区分这几种不同类型的畸形病状。注意观察有无病征。

2. 病毒内含体检查

①撕片法直接观察：取症状明显的烟草花叶病毒病的新鲜叶片，用刀片在叶背面的叶脉上切一个小口，用尖头镊子从花叶部位撕取一层下表皮，置于载玻片上的小水滴中，盖上盖玻片，在显微镜下直接观察。TMV病毒在表皮毛的基部细胞中常可见到晶体状内含体，多数为六角形。以健康叶片为对照，进行同样操作，观察是否有内含体。

②染色法：取感染蚕豆花叶病毒病症状明显的新鲜叶片，用镊子撕取病叶下表皮，放置于载玻片上，滴加鲁戈尔碘液染色2~3 min，加盖玻片，置于显微镜下观察。蚕豆表皮

细胞内的细胞核会被染成鲜黄色，而蚕豆花叶病毒的不定形内含体则被染成黄褐色。如用锥虫蓝染液，则细胞核在 30 s 内被染成蓝色，而病毒内含体则被染成深浅不匀的颗粒体。如用焰红染液，细胞核被染为粉红色，而病毒内含体则被染为鲜红色。以健康叶片为对照，进行同样操作，分析观察结果的差异。

【结果和讨论】

1. 分析植物病毒内含体的本质。
2. 拍摄本实验中观察到的病毒内含体的显微形态图。
3. 谈谈进行植物病毒内含体显微观察的注意事项。

EXPERIMENT 17　Observation of Symptoms and Detection of Inclusion Body of Plant Viral Diseases

【Introduction】

The main basis for diagnosis of plant viral diseases is the characteristics of symptom, mainly including discoloration, necrosis and malformation. The signs of plant virial diseases can't be observed on the surface of host tissues. However, for some plants, when they are infected by viruses, some crystalline or amorphous structures can be formed in the infected tissues, i. e. inclusion body. Plant viruses of different genera can produce different shaped inclusion bodies, such as amorphous, crystalline or pinwheel, and so on. The crystalline inclusion body is generally composed of virions arranged neatly and stacked, while the amorphous inclusion body is composed of virions and host cell components. The inclusion bodies existing in the cell nucleus are called intranuclear inclusion bodies, while those exist in the cytoplasm are called cytoplasmic inclusion bodies. Most of inclusion bodies are usually found in the plant tissues showing obvious symptoms. The large inclusion bodies can be observed under a light microscope, while the small inclusion bodies can just be observed under an electron microscope. The morphological difference of inclusion body can be taken as an important basis for the diagnosis of plant virus diseases or the identification of viruses. For example, only the viruses of *Potyvirus* can form the pinwheel inclusion body, while the viruses of *Tobamovirus* produce the crystalline plate-like inclusion body.

【Experimental Purpose】

1. Recognize the symptom characteristics and main symptom types of plant viral diseases.

2. Master the essence of the viral inclusion body and the observation technique of preparing slides by tear-off method.

【Materials and Apparatus】

1. Materials

Specimens of plant viral diseases: tobacco mosaic virus disease(TMV), broad bean mosaic virus disease (BBMV), cucumber mosaic virus disease (CMV), apple mosaic virus disease (AMV), soil-borne wheat mosaicvirus disease (SbWMV), tobacco ringspot virus disease (TRSV), tomato stripe virus disease(TSV), potato leaf roll virus disease(PLRV), tomato fern leaf virus disease(TFLV), and barley yellow dwarf virus disease(BYDV).

2. Reagents

①Lugol's iodine solution: Take 1.0 g iodine and 2.0 g potassium iodide, and place them in a mortar to grind. Add a certain amount of sterile water to the mortar slowly when grinding until iodine and potassium iodide dissolve. Next, the solution is transferred into a volumetric flask, and the sterile water is added to the constant volume of 300 mL. The solution is thoroughly mixed and then transferred to a brown reagent bottle, which is stored in the dark for later use.

②Trypan blue solution: Take 0.5 g trypan blue, and dissolve it in 100 mL heated 0.9 mol/L NaOH solution, which is considered as an stock solution and diluted 2 000-5 000 times when used.

③Flame red: It is also known as phloxine. Take 1 g phloxine to dissolve in 100 mL of water.

3. Instruments and Appliances

Microscopes, glass slides, coverslips, pointed tweezers, blades, floating agents, etc.

【Methods and Procedures】

1. Symptom Observation of Plant Viral Diseases

①Discoloration: Observe the specimens of tobacco mosaic virus disease, broad bean mosaic virus disease, cucumber mosaic virus disease, apple mosaic virus disease, and wheat soil-borne mosaic disease, and record their symptom characteristics, such as the features of leaf discoloration, the thickness of leaf and the phenomenon of fern leaf. Observe whether there are differences on the discoloration caused by the above plant viral diseases and there are signs or not.

②Necrosis: Observe the specimens of tobacco ringspot virus disease and tomato stripe virus disease, and record the characteristics of the two necrotic lesions and whether there are signs or not.

③Malformation: Observe the specimens of potato leaf roll virus disease, tomato fern leaf virus disease and barley yellow dwarf disease, and record their symptom characteristics. Note the morphological difference between the diseased and healthy tissues. Distinguish the different types of malformation and note whether there are signs or not.

2. Detection of Viral Inclusion Body

①Direct observation using a tear-off method: Take a fresh leaf with obvious symptoms of tobacco mosaic disease, cut a small cut on the vein on the back of the leaf with a blade, and then tear a layer of epidermis from the mosaic part using a pointed tweezer. Following it, the teared epidermis is placed in a small water droplet on a glass slide, covered by a coverslip, and directly observed under a microscope. The crystalline inclusion bodies can usually be seen for TMV in the basal cells of trichome, most of which are hexagonal. Take healthy leaves as a control and perform the same operation to observe whether there are inclusion bodies.

②Dyeing method: Take a fresh leaf with obvious symptoms of broad bean mosaic disease, tear off its back epidermis with a pointed tweezer, and then place it on a glass slide. Add one drop of Lugol's iodine solution to the slide for a staining for 2-3 min, cover it with a coverslip, and observe it under a microscope. The cell nuclei in the epidermal cells of broad bean are stained bright yellow, while the amorphous inclusion bodies of broad bean mosaic virus are stained yellowish brown. If the staining is carried out using trypan blue, the cell nuclei will be stained as blue within 30 s, while the viral inclusion bodies will be stained as uneven particles. If the staining is conducted with flame red dye, the cell nucleus will be stained pink while the viral inclusion bodies will be stained bright red. Healthy leaves are used as control to perform the same operation and analyze the differences on observation results.

【Results and Discussion】

1. Analyze the essence of viral inclusion bodies.
2. Photograph the microscopic morphology of the viral inclusion bodies observed in this experiment.
3. Talk about the precautions to observe viral inclusion bodies using a microscope.

实验十八 植物病毒的接种与传染

【概述】

植物病毒是一类核蛋白分子，无主动侵染的能力，但可经汁液摩擦(又称机械摩擦)、生物介体及嫁接进行传染。汁液摩擦传染常只发生于有病毒存在的表皮薄壁细胞内，即花叶型病毒，其机制是通过摩擦在健康寄主植物表皮薄壁细胞细胞壁和蜡质层上造成微伤口，病毒经微伤口进入细胞内而引起植物发病。

介体传染是大多数植物病毒田间扩散的重要方式。能传播病毒的介体很多，以刺吸式口器的昆虫为主，其次是真菌等。昆虫介体在含有病毒的植株上取食后，病毒可进入其口针和食道，甚至进入体液循环系统。当其再迁移到其他健康植株上取食时，就会把病毒传染到健康植株上。介体传染具有专化性，而且不同的介体与病毒组合的传播过程也不尽相同。嫁接传染实质上是病毒在愈合的病健组织之间进行传导的结果，只要砧木或接穗带毒，嫁接成功后，系统侵染的病毒就会使嫁接苗全株带毒。嫁接可以传播任何种类的病毒及类病毒。

【实验目的】

1. 掌握通过人工摩擦进行病毒汁液接种的关键技术。
2. 掌握介体接种植物病毒的操作技能。

【材料和器具】

1. 实验材料

烟草花叶病毒(TMV)；心叶烟的健康幼苗；蚜虫(不带任何病毒的健康无翅蚜)。

2. 实验器具

研钵和纱布(肥皂水清洗后，煮沸 15 min)、金刚砂(400~600 目)、洗瓶、养虫室、植物温室、培养箱、培养皿、毛笔、10%的烯啶虫胺 1 000 倍稀释液、1%的 K_2HPO_4缓冲液、离心机、离心管、标牌、肥皂等。

【方法和步骤】

1. 汁液摩擦接种

①病毒汁液提取：采集感染烟草花叶病毒症状典型的新鲜叶片，去除主脉后剪成小块，放入研钵中，加入适量的磷酸缓冲液，将叶片研磨成匀浆。然后用双层纱布对其过滤，收集滤液，滤液在 3 000 r/min 离心 30 min，收集上清液，为 TMV 病毒的粗提液，备用。

②摩擦接种：取健康心叶烟幼苗，在叶片上面撒一薄层金刚砂，用手指蘸取 TMV 的粗提液(双手须用肥皂洗净，并冲走肥皂液)，在撒有金刚砂的叶面上轻轻摩擦 2 次，然后立即用洗瓶冲洗叶片。取健康幼苗，以磷酸缓冲液代替 TMV 粗提液重复上述操作，作为对照。

③培养及症状观察：将接种后的所有植株放在植物温室中培养，培养温度为 25℃，相对湿度为 75%，光周期为 12 h/d，光照强度为 2 000 lx。每 2 d 观察记录一次实验结果，注意观察接种叶和新叶上的症状是否相同。

2. 蚜虫传毒接种

①蚜虫收集及处理：用振落法从隔离的养虫笼中收集无翅蚜，使蚜虫跌落到大培养皿中，用毛笔蘸取蚜虫收集在三角瓶中，纱布封口，在 20℃左右的温箱中使其饥饿 4~6 h。

②蚜虫饲毒：将烟草花叶病的新鲜病叶摘下放在大培养皿中，用毛笔取出饥饿的蚜虫，将其抖落在病叶上取食。

③蚜虫传毒：待蚜虫在病株上取食 10~15 min 后，用毛笔轻触蚜虫尾部使蚜虫拔出口针后，将带毒蚜虫转移至健康的烟草植株的心叶上，用养虫罩罩住。24 h 后喷洒 10%的烯啶虫胺 1 000 倍稀释液杀死蚜虫，移去养虫罩。

④培养及症状观察：将接种后的所有植株放在植物温室中培养，培养温度为 25℃ ，相对湿度为 75%，光周期为 12 h/d，光照强度为 2 000 lx。每 2 d 观察记录一次实验结果，注意将其接种叶和新叶上的症状是否相同。

【结果和讨论】

1. 详细记录并分析汁液摩擦接种和蚜虫传毒接种的实验结果。
2. 为什么摩擦接种时要在接种叶面撒放金刚砂？
3. 请分析汁液摩擦接种前必须洗手的原因。
4. 接种植物病毒还可以采用哪些方法？如何选择合适的方法接种植物病毒？

EXPERIMENT 18　Inoculation and Infection of Plant Virus

【Introduction】

Plant virus is a type of nucleoprotein molecule that has no ability to infect actively but can infect into plants by juice fluid friction (also known as mechanical friction), biological vectors, and grafting. The infection of juice fluid friction usually only occurs in epidermal parenchyma cells where the virus is present, i. e, mosaic virus. Its mechanism is to create micro wounds on the cell wall and waxy layer of healthy host plant epidermal parenchyma cells through friction, and the viruses can enter cells through micro wounds to cause plant diseases.

Vector is an important medium for most plant viruses to spread in the field. There are many vectors that can transmit viruses, most of which are insects with piercing-sucking mouthparts, followed by fungi. After the insect vector feeds on the plant containing viruses, the viruses can enter its stylet, esophagus, and even the circulation system of body fluid. When the insect migrates to other healthy plants for food, the viruses contained in the insects will be transmitted to the healthy plants. The infection through vectors shows specialization, and the transmitting pathway is different varied with the combination of vectors and viruses. Grafting infection is essentially the result that viruses are transmitted between the healed diseased and healthy tissues. As long as the rootstock or scion carries viruses, the whole grafted seedlings will be infected for the systematical infection viruses after the grafting is successful. Grafting can transmit any kind of virus and viroid.

【Experimental Purpose】

1. Master the key technology of viral juice inoculation by artificial friction.
2. Master the operation skills of vector inoculation for plant viruses.

【Materials and Apparatus】

1. Materials

Tobacco mosaic virus(TMV), healthy seedlings of *Nicotiana glutinosa*, and aphids(healthy wingless aphids without any virus).

2. Instruments and Appliances

Mortar and gauze that are washed by soap water and then autoclaved for 15 min, emery (400-600 mesh), washing bottles, insectary, plant greenhouse, incubators, Petri dishes, writing brushs, 1 000 times dilution of 10% nitenpyram, 1% K_2HPO_4 buffer, centrifuge, centrifuge tubes, labels, soap, etc.

【Methods and Procedures】

1. Juice Friction Inoculation

①Extraction of viral juice: Collect the fresh diseased leaves with obvious symptom of tobacco mosaic, cut them into small pieces after the main veins are removed, and then put them in a mortar for grinding into a homogenate in phosphate buffer. The homogenate is filtered by double gauze. The filtrate is collected and centrifuged at 3 000 r/min for 30 min. The supernatant is collected, i. e. the crude extract of TMV juice, ready for later inoculation.

②Friction inoculation: Take the healthy tobacco seedlings, sprinkle a thin layer of emery on the leaves, dip your finger into the crude extract of TMV(hands must be washed with soap and washed away with the soap), and then rub the leaves sprinkled with emery lightly for twice. Next, the leaves are immediately rinsed with a washing bottle. The healthy seedlings subjected to the above operations with phosphate buffer instead of TMV crude extract are taken as control.

③Cultivation and symptom observation: All inoculated plants are cultured in a greenhouse under a temperature of 25℃, a relative humidity of 75%, and a photoperiod of 12 h/d with a light intensity of 2 000 lx. Observe and record the experimental results at intervals of 2 d, noting whether the symptoms on the inoculated leaves are the same as those of newly budded leaves.

2. Aphid Feeding Inoculation

①Collection and treatment of aphids: The wingless aphids in an isolated insect cage are collected using the shaking method to make the aphids fall into a large Petri dish. The aphids are then collected into a conical flask using a brush. The flask is sealed with gauze and then kept in an incubator at 20℃ for a starvation of 4-6 h.

②Virus acquisition by aphids: Collect fresh diseased tobacco mosaic leaves and place them in a large Petri dish. Take out the starved aphids and shake them off on the diseased leaves for feeding.

③Virus transmission by aphids: After the aphids feed on the diseased plant for 10-15 min, use a brush to touch the tail of the aphid lightly with a brush to pull out the stylet. Next, the aphids are transferred to the healthy plant of *N. glutinosa* to be inoculated, which is covered with an insect hood. After 24 h, spray 1 000 times dilution of 10% nitenpyram to kill the aphids and remove the insect hood.

④Cultivation and symptom observation: All inoculated plants are cultured in a greenhouse under a temperature of 25℃, a relative humidity of 75%, and a photoperiod of 12 h/d with a light intensity of 2 000 lx. Observe and record the experimental results at intervals of 2 d, noting whether the symptoms on the inoculated leaves are the same as those of newly budded leaves.

【Results and Discussion】

1. Record the experimental results of juice friction inoculation and aphid feeding inoculation in detail.
2. Why is emery sprinkled on the inoculated leaves during friction inoculation?
3. Analyze the reasons of washing hands before juice friction inoculation.
4. What are the other methods that can be used for the inoculation of plant viruses? How to choose a suitable method to inoculate plant viruses?

实验十九 植物线虫的分离及形态观察

【概述】

线虫是一种低等动物，属于无脊椎动物线形动物门的线虫纲。它的种类很多，少部分寄生在植物体内引起植物病害，为害植物根部会引起根结、根肿等症状，地上部分生长不良；

为害茎造成萎蔫、枯死等症状；为害叶会造成叶斑、小叶、皱缩和黄化等症状；为害果实和种子产生虫瘿和肿瘤等畸形症状。与植物病害相关的线虫主要有垫刃目(Tylenchida)、滑刃目(Aphelenchida)、矛线目(Dorylaimida)及三矛目(Triplonchida)。

植物寄生线虫的消化系统和生殖系统发达，其次是神经系统，其他系统则退化。消化系统包括口、口腔、食道、肠、肛门。食道类型是植物线虫分目的最主要依据，植物病原线虫的口腔内均有口针。雌虫的生殖系统包括卵巢(单个或成双)、输卵管、受精囊、子宫、阴道、阴门。雄虫的生殖系统包括精巢(单个或成双)、输精管和泄殖腔。泄殖腔内有一对交合刺，有的还有引带和交合伞等附属器官。雄虫的生殖孔和肛门是同一个孔口，称为泄殖孔。

植物寄生线虫个体很小，除极少数可从植物组织中直接挑出外，绝大多数需要借助特定的工具和方法才能分离出来。线虫的分离主要是利用它的趋水性及大小和密度与其他杂质的差异，采用过筛、离心或漂浮等措施，将线虫从植物组织或土壤中分离出来。

【实验目的】

1. 了解植物病原线虫所致病害的症状特点。
2. 掌握植物线虫的分离方法，识别主要植物病原线虫的形态与显微特征。
3. 识别并掌握植物病原线虫食道的类型及特点。

【材料和器具】

1. 实验材料

①植物线虫病害标本：花椒根结线虫病、松材线虫病标本、大豆胞囊线虫病、小麦粒线虫病、水稻干尖线虫病、花生根结线虫病等。

②植物病害组织和土壤：松材线虫病的病木、花椒根结线虫病的根结及根际土壤。

③TAF 固定液：取三乙醇胺 2 mL 和 40%甲醇 7 mL，将其溶解在 91 mL 的蒸馏水中。

2. 实验器具

显微镜、解剖镜、天平、剪刀、锯子、剪枝剪、漏斗架、带橡胶管的玻璃漏斗、止水夹、筛子(60 目、100 目、200 目和 325 目)、竹签、小培养皿、塑料桶、烧杯、载玻片、盖玻片、胶头滴管、纱布、吸水纸等。

【方法和步骤】

1. 植物线虫病的症状观察

观察并记录供试植物线虫病害的症状，比较不同植物线虫病害所致症状的差异。

2. 线虫的分离及形态观察

(1)漏斗分离法

该方法常用来分离植物组织和土壤中能运动的线虫，操作简单，利用线虫的趋水性和自身质量的特点，使其脱离组织或土壤，并游离到水中集中沉降在橡皮管的下端，从而将线虫收集。具体操作步骤如下：

①将干净的漏斗放在漏斗架上，其底部橡胶管末端用止水夹夹住，向漏斗中加入 200 mL 左右的自来水。

②取花椒根结线虫病的根结用剪刀剪碎，也可取松材线虫病的病木先用锯子锯开，再用剪枝剪剪成小块。取剪碎的病组织材料用双层纱布包好，轻放入盛有水的漏斗中，然后向漏斗中添加适量的水将其补满，静置 24 h。

③轻轻打开止水夹，向小培养皿中放入约 5 mL 液体。

④将小培养皿置于解剖镜下，用竹签(或用微型吸管)挑取线虫，移至载玻片上的 TAF 固定液滴中，加盖玻片。

⑤将载玻片在酒精灯火焰上来回移动 3~4 次，通过加热处理使线虫失活，当弯曲的线虫突然变为僵直状，说明线虫已被杀死，应立即停止加热。

⑥将载玻片放置在显微镜下，先在低倍镜下观察线虫的整体形态结构，然后在高倍镜下重点观察线虫的食道类型与生殖系统结构。

(2)过筛分离法

此方法适合分离土壤中运动和不运动的线虫。利用线虫的趋水性，首先使线虫悬浮在水中，再用筛网将线虫与土壤分开，最后从水中分离线虫。具体操作步骤如下：

①将花椒根结线虫病株根际土壤(100~150 g)放入装有适量水的塑料桶 A 内，充分搅拌至土块碎散，静置 30 s。

②静置后，收集塑料桶 A 中的上层悬浮液，过 60 目的筛子，将滤液收集在干净的塑料桶 B 内。

③将塑料桶 B 中的液体按顺序依次通过 100 目、200 目和 325 目的筛子，3 个筛子上的残留物分别洗脱在 3 个烧杯中。

④吸取 325 目筛子的洗脱液于小培养皿内，置于解剖镜下，用竹签(或用微型吸管)挑取线虫，移至载玻片上的 TAF 固定液滴中，加盖玻片。

注意：若线虫数量太少，可将洗脱液倒入离心管中，1 500 r/min 离心 3 min 后，弃上清液，将下层沉淀物用少量的水重新悬浮后，再镜检。

⑤将载玻片放置在显微镜下，先在低倍镜下观察线虫的整体形态结构，然后在高倍镜下重点观察线虫的食道类型与生殖系统结构。

【结果和讨论】

1. 线虫引起的植物病害症状有哪些特点？

2. 绘制分离到的植物病原线虫的形态特征及显微结构图，并对食道及生殖系统的构造进行标注。

3. 寄生性线虫与腐生性线虫有哪些不同？在镜检时如何区分？

EXPERIMENT 19　Isolation and Morphological Observation of Plant Nematode

【Introduction】

Nematode is a type of lower animal belonging to the class of Nematoda in the phylum of Nematomorpha in invertebrates. There are many kinds of nematodes, and a small part of nematodes can parasitize in plants and cause plant diseases. Nematodes can damage roots to cause root knots and root swelling and weak growth of the above-ground parts, damage stems to cause wilting and death, damage leaves to cause leaf spots, leaflets, wrinkles, and yellowing, and damage fruits and seeds to result in malformation such as galls and tumors. The main nematodes related to plant diseases are the orders of Tylenchida, Aphelenchida, Dorylaimida and Triplonchida.

The digestive and reproductive systems of plant parasitic nematodes are highly developed, followed by the nervous system, and the others are degraded. The digestive system includes mouth, mouth cavity, esophagus, intestine, and anus. The type of esophagus is the main criteria for the classification of plant nematodes on the level of order. There is stylet in the mouth cavity for all plant pathogenic nematodes. The reproductive system of female nematode includes ovary(single or double), fallopian tube, spermathecal, uterus, vagina, and vulva, while that of male nematode includes testis(single or double), spermaduct and cloaca. There is a pair of mating spicules in the cloaca, and sometimes there are the other appendages such as gubernaculum and copulatory bursa. The gonopore and anus of male nematode are the same orifice, which is named as cloacal pore.

The size of plant-parasitic nematode is extremely small. Except for a few that can be picked directly from plant tissues, it is necessary to employ specific tools and methods to isolate most nematodes. The separation of nematode is mainly based on its hydrotaxis and the differences in size and density between nematodes and other impurities. The nematodes can be isolated from plant tissues or soil by the measures of sieving, centrifugation or floating.

【Experimental Purpose】

1. Understand the symptom characteristics of plant nematode diseases.

2. Master the isolation methods of plant nematodes, and recognize the morphological and microscopic features of main plant pathogenic nematodes.

3. Identify and master the types and features of esophagus of plant pathogenic nematodes.

【Materials and Apparatus】

1. Materials

①Plant nematode specimens: pine wood nematode, prickly ash root knot, soybean cyst nematode, wheat seed-gall nematode, rice white-tip nematode and peanut root knot nematode, etc.

②Plant diseased tissues and soil: the diseased stems of pine wilt nematode, and the root knot and rhizosphere soil of prickly ash root knots.

③TAF fixative: Take 2 mL triethanolamine and 7 mL 40% methanol, and dissolve them in 91 mL distilled water.

2. Instruments and Appliances

Microscopes, dissecting microscopes, balances, scissors, saws, pruning shears, funnel stands, glass funnels with rubber tubes, pinchcocks, sieves(60-mesh, 100-mesh, 200-mesh, 325-

mesh), bamboo skewers, small Petri dishes, plastic buckets, beakers, glass slides, coverslips, rubber droppers, gauzes, absorbent papers, etc.

【Methods and Procedures】

1. Symptom Observation of Plant Nematode Diseases

Observe and record the symptoms of all provided plant nematode diseases, and compare the differences in symptoms of different plant nematode diseases.

2. Isolation and Morphological Observation of Nematodes

(1) Funnel Isolation

This method, which is usually employed to isolate the motile nematodes in plant tissues and soils, is extremely simple to operate. It utilizes the hydrotaxis and their own quanlity characteristics of nematodes to make them detach from the tissues and soils. The nematodes can swim into the water and deposit at the bottom of the rubber tube to be collected. The specific operation steps are as follows:

①Place a clean funnel on the funnel rack, clamp the end of the rubber tube at the bottom with a pinchcock, and add about 200 mL of tap water to the funnel.

②Cut the root knot of pepper root knot nematode with scissors, or saw the diseased wood of pine wood nematode and then cut into small pieces with pruning shears. Wrap the cut pieces of diseased tissue material with a double gauze, put it into a funnel filled with water gently, add a certain amount of water to fill in the funnel, and then let it stand for 24.

③Open the pinchcock gently, and collect 5 mL liquid using a small Petri dish.

④Place the small Petri dish under the dissecting microscope, pick up the nematodes with a bamboo skewer (or use a micropipette), transfer them into the TAF fixative on a glass slide and covered with a coverslip.

⑤The slide is moved back and forth on the flame of an alcohol lamp for 3-4 times, which can make nematodes inactivate by heating treatment. When the bent nematodes suddenly become rigid, it indicates that the nematodes have been killed, and the heating should be stopped immediately.

⑥Place the slide under a microscope, observe the overall morphological structure of the nematodes under a low magnification firstly, and then observe the types of esophagus and the structure of the reproductive system of the nematode under a high magnification.

(2) Sieving Isolation

This method is suitable for isolating motile and non-motile nematodes in soil. Based on the hydrotaxis of the nematodes, they are suspended in water fistly, then separated from the soil with a sieve, and finally the nematodes are isolated from the water. The specific operation procedures are as follows:

①Put the rhizosphere soil of the diseased seedlings of prickly ash root knot nematode (100-150 g) into a plastic bucket A with an appropriate amount of water, stir thoroughly until the soil is broken up, and let it stand for 30 s.

②Collect the upper suspension in the plastic bucket A after standing, filter it through a 60-mesh sieve, and collect the filtrate in a clean plastic bucket B.

③Filter the liquid in the plastic bucket B through the sieves of 100-mesh, 200-mesh, and 325-mesh in sequence, and the residues on the three above sieves are eluted respectively into three different beakers.

④Take the eluate from the 325-mesh sieve into a small Petri dish, place it under a dissecting microscope, pick up the nematodes with a bamboo skewer (or use a micropipette), and transfer them into the TAF fixative on a glass slide and covered with a coverslip.

Note: If the number of nematodes in the eluate is too few, you can pour the eluate into a centrifuge tube, centrifuge it at 1 500 r/min for 3 min, discard the supernatant, resuspend the sediment in lower layer with a small amount of water, and then check it under a microscope.

⑤Place the slide under a microscope, observe the overall morphological structure of the nematodes under a low magnification firstly, and then observe the types of esophagus and the structure of the reproductive system of the nematode under a high magnification.

【Results and Discussion】

1. What are the characteristics of plant disease symptoms caused by nematodes?

2. Draw the illustrations of morphological features and microscopic structures of the isolated plant pathogenic nematodes, and mark the structures of esophagus and reproductive systems.

3. What are the differences between parasitic nematodes and saprophytic nematodes? How do you distinguish them under a microscope?

实验二十　寄生性植物及植物寄生螨类的形态观察

【概述】

高等种子植物营寄生生活的并不多，分布在6~7个科，且多半都是双子叶植物，寄生在植物的地上部分或根部。根据它们对寄主的依赖程度可分为全寄生和半寄生，与木本植物有关的主要有桑寄生科(Loranthaceae)和菟丝子科(Cuscutaceae)。

螨类是一些小型或微小、肉眼几乎不能看见的动物，属于蛛形纲中的蜱螨目(Acarina)。寄生植物的螨类主要为叶瘿螨科(Eriophyidae)，这些螨类危害多种阔叶树和农作物的叶片或果实，受害部位因受刺激而畸形，形成虫瘿或导致毛毡病。

【实验目的】

1. 认识常见寄生性种子植物(属)的形态特征。

2. 观察植物寄生性螨类的基本形态及其所致病害症状。

【材料和器具】

1. 实验材料

桑寄生、槲寄生和矮槲寄生的盒装标本；菟丝子的新鲜或干标本(带寄主茎)；葡萄毛毡病(或槭树毛毡病)的新鲜标本。

2. 实验器具

显微镜、通草、培养皿、刀片、蒸馏水、载玻片、盖玻片、纱布等。

【方法和步骤】

1. 桑寄生科重要属识别

属半寄生性植物，具叶绿体，能进行正常的光合作用，根退化为吸根，伸入寄主木质部并与其导管相连，从寄主体内获取矿物质、水分和生长物质等。

观察以下3个属的代表种标本，区别它们的形态特征。

①桑寄生属(*Loranthus*)：茎寄生。常绿小灌木、叶互生或近对生，羽状叶脉明显；枝条圆柱形、被星状毛；花两性、形大，浆果。如欧洲桑寄生(*L. europaeus*)。

②槲寄生属(*Viscum*)：常绿小灌木，枝叶均为绿色；二叉状分枝；叶对生，叶脉不明显三出；花两性或单性、形小；浆果、半透明。如槲寄生(*V. album*)。

③矮槲寄生属(*Arceuthobium*)：植株高3~8 cm，枝条对生，分枝多而成丛，叶退化为鳞片状，绿色或黄绿色。如云杉矮槲寄生(*A. sichuanense*)。

2. 菟丝子科重要属识别

菟丝子属(*Cuscuta*)：属全寄生性植物，为缠绕茎的草本植物，叶退化为鳞片状，茎叶呈黄色、橙黄色或紫褐色，花常为白色。

取被菟丝子寄生的植物茎，先观察茎干处缠绕的菟丝子形态，再以寄主的茎(应为菟丝子吸根伸入的部位)做徒手切片，观察菟丝子吸根与寄主组织间的关系。

3. 植物寄生螨的形态观察

叶瘿螨属(*Eriophyes*)：身体蠕虫形、狭长。极微小，长约0.1 mm，足2对，前肢体段背板成盾状，后肢体段延长，分为很多环纹，无呼吸系统。

取葡萄毛毡病(*E. vitis*)或槭树毛毡病叶片，首先观察叶子的畸形病斑，并注意叶背病斑的毛毡状物，再用带有毛毡的病斑做徒手切片，镜下观察寄主畸形表皮细胞和螨的形态。

【结果和讨论】

1. 比较观察到的寄生性植物的形态特征及寄生性。
2. 绘制菟丝子吸根在寄主组织内的扩展图。
3. 绘制螨在植物叶上寄生的形态图。
4. 被寄生性种子植物寄生的寄主植物会不会枯死？为什么？
5. 植物寄生在植物上，这是生物的进化还是退化？

EXPERIMENT 20　Morphological Observations of Parasitic Plants and Mites

【Introduction】

There are not many higher seed plants that live in parasitism. They are distributed in 6-7 families, and most of them are dicotyledonous plants, parasitizing on the aboveground parts or roots of plants. According to the degree of their dependence on the host, they can be divided into full parasitic and semi-parasitic plants. Parasitizing plants related to woody plants are mainly Loranthaceae and Cuscutaceae.

Mites are small or tiny animals that are almost invisible to the naked eye and belong to the order of Acarina in the class of Arachnoidea. The parasitic mites on plants mainly include the family Eriophyidae. These mites damage the leaves or fruits of a variety of broad-leaved trees and crops. The affected parts are malformation due to stimulation, causing gall mite or erineum mite diseases.

【Experimental Purpose】

1. Understand the morphological characteristics of common parasitic seed plants(genus).

2. Observe the basic morphology of plant-parasitic mites and the disease symptoms caused by them.

【Materials and Apparatus】

1. Materials

Boxed specimens of *Loranthus*, *Viscum*, and *Arceuthobium*; fresh or dried specimens of *Cuscuta*(with host stem); fresh specimens of grape erineum mite disease(or maple erineum mite disease).

2. Instruments and Appliances

Microscopes, ricepaperplant piths, Petri dishes, blades, distilled water, glass slides, coverslips, gauzes, etc.

【Methods and Procedures】

1. Identification of Important Genera in Loranthaceae

The Loranthaceae plants are semiparasitic, which have chloroplasts and are able to proceed normal photosynthesis. The roots of them degenerate into haustoria, which extend into the xylem of the host and connect to its ducts, and obtain minerals, water and growth materials from the host.

Observe the representative specimens of the following three genera to distinguish their morphological characteristics.

①*Loranthus*: It parasitizes in stems. They are mall evergreen shrubs with alternate or nearly opposite leaves and obvious pinnate leaf veins. The branches are cylindrical covered by stellate seta. The flowers are bisexual with large shape, and the fruits are berries, such as *L. europaeus*.

②*Viscum*: It belongs to small evergreen shrubs with green branches and leaves. The branches are bifurcated and the leaves grow in opposite with not obvious veins. The flowers are bisexual or unisexual with small shape. The fruits are translucent berries, such as *V. album*.

③*Arceuthobium*: The plant is 3-8 cm high with opposite branches. The branches are multiple to form clumps, and the leaves degenerate into scales with green or yellow-green in color, such as *A. sichuanense*.

2. Identification of Important Genera in Cuscutaceae

Cuscuta: It is holoparasitic herb with twining stems, its leaves are reduced to scales, its stems and leaves are yellow, orange or purple-brown and its flowers are often white.

Take the plant stem parasitized by *Cuscuta*, first observe the morphology of *Cuscuta* entangled at the stem, and then take the stem of the host (which should be the part where the haustoria penetrate) to make a section to observe the relationship between the haustoria of *Cuscuta* and the host tissue.

3. Morphological Observation of Plant-parasitic Mites

Eriophyes: The body is worm-shaped, narrow, and long. It is very small, about 0.1 mm in length, with 2 pairs of feet. The back of the forelimb segment is shield-shaped. The hindlimb segment is extended, which is divided into many rings. It has no respiratory system.

Take grape or mapleerineum mites (*E. vitis*) leaves, observe the malformation of the leaves, and pay attention to the lesion with the erineum mites on the back of the leaves, then make handworked sections of the lesions, and observe the morphology of the host malformed epidermal cells and mites under a microscope.

【Results and Discussion】

1. Compare the morphological characteristics and parasitism of the parasitic plants observed.
2. Draw a picture of the haustoria extension of *Cuscuta* in the host tissues.
3. Draw a morphological illustration of the parasitic mites on plant leaves.
4. Will the host plants that are parasitized by parasitic seed plants die?
5. Is the phenomenon that plants parasitize plants evolutionary or degenerated for organisms?

实验二十一　植物病原真菌孢子的诱导产生和萌发

【概述】

真菌孢子在植物病害循环中具有重要的作用，其既可以作为初侵染来源，也可以作为病害传播与再侵染的重要来源。植物病原真菌孢子的产生和萌发是植物真菌病害研究的重要问题。明确真菌孢子的产生和萌发条件是非常必要的，能为病原真菌的生活史与侵染过程、植物病害的发病规律及杀菌剂药效测定等方面的研究提供重要指导依据。人工培养的真菌，其孢子的产生和萌发受内在条件和外界条件的影响。

【实验目的】

1. 学习并掌握真菌产孢的诱导技术。
2. 掌握真菌孢子萌发的测定方法。
3. 了解环境条件对真菌孢子产生和萌发的影响。

【材料和器具】

1. 实验材料

(1)植物病原真菌

花椒干腐病(*Fusarium zanthoxyli*)和侧柏叶枯病(*Alternaria alternata*)。

(2)培养基

①PDA 培养基：将马铃薯洗净去皮，切成小块，称取 200 g，放入锅中，加水 1 000 mL，煮沸 30 min，稍冷却后用纱布过滤，滤液中加入葡萄糖 20 g 和经加热融化的琼脂 15~20 g，搅拌均匀，加水补充到 1 000 mL，趁热分装在三角瓶中，高压灭菌后备用。

②花椒汁液培养基(Zb-PDA)：采集新鲜健康的花椒嫩枝 5 g，剪碎后加少量无菌水研磨，用无菌水定容至 1 000 mL，以 3 000 r/min 离心 20 min，收集上清液，用此花椒汁液来配制 PDA，高压灭菌后备用。

③1/2 胡萝卜培养基(1/2CA)：100 g 胡萝卜加入 450 mL 水中，用榨汁机制备成匀浆状，将其转移至三角瓶中，加入 10 g 琼脂粉，高压灭菌后备用。

(3)营养液

①0.3%侧柏汁液：采集新鲜健康的侧柏叶片 0.3 g，加少量无菌水研磨捣碎，用无菌水定容至 100 mL，以 3 000 r/min 离心 20 min，收集上清液，将其 pH 值调至 7.0，过滤除菌后备用。

②0.3%葡萄糖液：称取 0.3 g 葡萄糖，将其溶解于 100 mL 无菌水中，将其 pH 值调至 7.0，过滤除菌后备用。

③0.3%蔗糖液：称取 0.3 g 蔗糖，将其溶解于 100 mL 无菌水中，将其 pH 值调至 7.0，过滤除菌后备用。

④0.3%马铃薯汁液：称取去皮的土豆 0.3 g，加少量无菌水研磨捣碎，用无菌水定容至 100 mL，以 3 000 r/min 离心 20 min，收集上清液，将其 pH 值调至 7.0，过滤除菌后备用。

(4)试剂

0.1 mol/L NaOH 溶液、0.1 mol/L HCl 溶液、硫酸、无菌水等。

2. 实验器具

恒温培养箱、显微镜、打孔器、牙签、无菌培养皿、血球计数器、棉签、凹玻片、盖玻片、移液枪、枪头、吸水纸、pH 试纸、干燥器、离心机等。

【方法和步骤】

1. 真菌的产孢诱导

真菌孢子的产生与培养基的成分及培养条件相关。不同真菌孢子的产生对培养基的营养成分要求不同，如培养基的碳源和氮源、特殊微量元素、维生素、激素等。有些真菌在植物性培养基上能够更好地产生孢子。影响真菌产孢的培养条件主要包括光照、温度和通气量。要确定某种真菌产孢的最佳条件，需要对培养基的成分和培养条件进行实验。本实验以花椒干腐病菌病原为例，研究不同培养基和光照对孢子产生的影响。

①将保存于-80℃的花椒干腐病病原菌菌种取出，接种于 PDA 培养基平板上，25℃暗培养 7 d 后，备用。

②用无菌打孔器花椒在干腐病病原菌的菌落边缘上制备菌饼，菌饼直径约 5 mm，备用。

③用灭菌牙签挑取菌饼分别接种于 PDA、Zb-PDA 和 1/2 CA 培养基平板的中心，每种培养基接种 6 个平板，其中 3 个平板放置在光照培养箱中培养，另外 3 个避光培养。

④培养 14 d 后，向各平板中添加 10 mL 无菌水，用灭菌的棉签在菌落表面轻轻擦拭，洗下真菌孢子，用灭菌的双层纱布过滤，制备成孢子悬浮液。

⑤显微镜下观察花椒干腐病病原菌在这 3 种培养基上的产孢差异，如分生孢子的类型和分生孢子的形态(镰刀菌会产生镰刀型大型分生孢子和圆形小型分生孢子)，并用血球计数器计算孢子的浓度。

2. 环境条件对孢子萌发的影响

环境条件能影响孢子的萌发率、萌发速率及芽管的长度，甚至萌发方式。本实验以侧柏叶枯病病原菌为供试菌种，该病原产孢能力极强，能确保实验所需的孢子量。

①孢子悬浮液的制备：从冰箱中取出菌种，接种于 PDA 培养基平板上，25℃暗培养 7 d 后，添加 0.3%马铃薯汁液，用灭菌的棉签在菌落表面轻轻擦拭，洗下真菌孢子，用灭菌的双层纱布过滤，制备成孢子悬浮液。取一滴孢子悬浮液，显微镜下检查，调整孢子浓度至 1×10^6个/mL 左右，备用。

②孢子萌发率的计算：孢子萌发的标准是芽管长度大于孢子直径的 1/2，检查孢子总数不低于 300 个。萌发率计算公式为：萌发率(%)= 萌发孢子数×100/检查孢子总数。

③温度对孢子萌发的影响：用移液枪吸取 50 μL 孢子悬浮液滴加在凹玻片的凹槽内，盖上盖玻片，分别放在 5℃、10℃、15℃、20℃、25℃、30℃、35℃和 40℃温箱中，可每人做 2 个温度处理，每个处理 3 次重复。培养 12 h 后，显微镜下检查孢子萌发情况，计算每个温度下的孢子萌发率。

④湿度对孢子萌发的影响：用移液枪吸取 50 μL 孢子悬浮液滴加在凹玻片的凹槽内，盖上盖玻片，分别放在相对湿度为 50%、75%、90%和 100%的硫酸干燥器中，培养温度为 25℃，这 4 个湿度对应的硫酸浓度分别为 43.4%、30.0%、18.5%和 0，每个处理 3 个重复。培养 12 h 后，显微镜下检查孢子萌发情况，计算每个湿度下的孢子萌发率。

⑤pH 值对孢子萌发的影响：预先配置不同 pH 值的无菌水，pH 值分别为 3、4、5、6、7、8、9 和 10。取孢子悬浮液与不同 pH 值的无菌水等量混匀，用移液枪吸取 50 μL 该孢子悬浮液滴加在凹玻片的凹槽内，盖上盖玻片，放置在 25℃培养箱内培养 12 h 后，显微镜下观察孢子萌发情况，计算每个湿度下的孢子萌发率。

⑥营养成分对孢子萌发的影响：将孢子悬浮液等量分装在不同的离心管内，2 000 r/min 离心 3~5 min，弃上清液后，向离心管内添加无菌水重悬底部的孢子以清洗残留的培养基，离心后弃上清液，重复 2~3 次。向各离心管内分别添加等量的 0.3%侧柏汁液、0.3%葡萄糖液、0.3% 蔗糖溶液和 0.3%马铃薯汁液，并使孢子完全悬浮在营养液中，以添加等量无菌水为对照，每个处理 3 个重复。将各处理放置在 25℃振荡培养箱内暗培养 12 h 后，吸取各离心管内的孢子悬浮液，制作临时显微玻片，显微镜下检查孢子萌发情况，计算各营养液处理下的孢子萌发率。

【结果和讨论】

1. 计算不同诱导条件下花椒干腐病病原菌的产孢浓度，分析不同条件对孢子产生的影响。

2. 计算不同环境条件下侧柏叶枯病病原菌孢子的萌发率，分析环境条件对孢子萌发的影响。

3. 孢子萌发测定技术可以应用在植物病害研究的哪些方面?

4. 孢子在植物病程及病害循环中有什么作用?

EXPERIMENT 21　Induced Production and Germination of Spores of Plant Pathogenic Fungi

【Introduction】

Fungal spores play an important role in the cycle of plant diseases, which can be the source of primary infection as well as the source of disease transmission and re-infection. The production and germination of spores of plant pathogenic fungi are important issues in the study of fungal diseases. The induction of sporulation is an important content of numerous plant disease researches. It is very necessary to clarify the conditions of the production and germination of fungal spores, which can provide important guidance for the study of the life history and infection process of pathogenic fungi, the occurrence regularity of plant diseases and the determination of fungicide efficacy. The sporulation and spore germination of artificially cultured fungi are affected by internal and external conditions.

【Experimental Purpose】

1. Learn and master the method of fungal sporulation induction.
2. Master the method of determination of fungal spore germination.
3. Understand the influence of environmental conditions on fungal sporulation and spore germination.

【Materials and Apparatus】

1. Materials

(1) Phytopathogenic Fungi

Fusarium zanthoxyli and *Alternaria alternata*.

(2) Media

①Potato dextrose agar(PDA) medium: Wash and peel the potato, cut it into small pieces, weigh 200 g, put it into the pot, add 1 000 mL water, boil it for 30 min, after cooling slightly, filter it with gauze, add 20 g glucose and 15-20 g melted agar by heating into the filtrate, stir evenly, add water to 1 000 mL, and divide it into Erlenmeyer flasks while hot, which are then autoclaved for later use.

②*Zanthoxylum bungeanum* juice medium(Zb-PDA): Collect 5 g of fresh and healthy twigs of *Z. bungeanum*, cut and grind the twigs in a small amount of sterile water, dilute it to 1 000 mL with sterile water, centrifuge it at 3 000 r/min for 20 min, and collect the supernatant to prepare the PDA medium, which is then autoclaved for later use.

③1/2 Carrot medium (1/2 CA): Add 100 g carrots to 450 mL water, homogenate and transfer it to a Erlenmeyer flask, and add 10 g agar powder, which is then autoclaved for later use.

(3) Nutrient Solutions

①0.3% *Platycladus orientalis* juice: Collect 0.3 g of fresh and healthy leaves of

P. orientalis, and add a small amount of sterile water and grind them, which is then diluted to 100 mL with sterile water and centrifuge at 3 000 r/min for 20 min. The supernatant is collected and the pH is adjusted as 7.0, which is then subjected to filtration sterilization for later use.

②0.3% Glucose solution: Weigh 0.3 g glucose, which is then dissolved in 100 mL sterile water. The pH of the solution is adjusted as 7.0, after which, it is subjected to filtration sterilization for later use.

③0.3% Sucrose solution: Weigh 0.3 g sucrose, which is then dissolved in 100 mL sterile water. The pH of the solution is adjusted as 7.0, after which, it is subjected to filtration sterilization for later use.

④0.3% Potato juice: Weigh 0.3 g of peeled potato and grind it in a small amount of sterile water, which is then diluted to 100 mL with sterile water and centrifuged at 3 000 r/min for 20 min. The supernatant is collected and the pH is adjusted as 7.0, which is then subjected to filtration sterilization for later use.

(4) Reagents

Sodium hydroxide solution (0.1 mol/L), chlorhydric acid solution (0.1 mol/L), sulfuric acid, and sterile water, etc.

2. Instruments and Appliances

Constant temperature incubators, microscopes, hole punchers, toothpicks, sterile Petri dishes, hemocytometers, cotton swabs, concave glass slides, coverslips, pipettes, pipette tips, absorbent paper, pH test papers, dryer, centrifuges, etc.

【Methods and Procedures】

1. Induction of Fungal Sporulation

The fungal sporulation is related to the composition of the culture medium and the culture conditions. The sporulation of different fungi has different requirements on the nutrient composition of the medium, such as carbon source and nitrogen source of the medium, special trace elements, vitamins, and hormones. Some fungi can better produce spores on plant-based media. The culture conditions that affect fungal sporulation mainly include light, temperature, and ventilation. To determine the best conditions for the sporulation of a certain fungus, it is necessary to test the composition of the medium and the culture conditions. This experiment took *F. zanthoxyli* as an example to study the effects of different media and light on sporulation.

①Take the strain of *F. zanthoxyli* from −80℃, inoculate it to a PDA plate, and culture it in dark at 25℃ for 7 d.

②Prepare discs (5 mm in diameter) on the edge of the colony of *F. zanthoxyli* with a sterile puncher.

③Pick out the fungal discs using a sterilized toothpick and inoculate them to the center of PDA, Zb-PDA, and 1/2 CA plates. The cultures on each medium contains 6 plates, 3 of which are incubated in light, and the other 3 are incubated in the dark.

④After an incubation of 14 d, add 10 mL of sterile water to each plate, wipe gently on the surface of the colony with a sterilized cotton swab, wash the fungal spores off, and filter it through a sterile double gauze to prepare spore suspensions.

⑤Observe the type and morphology of conidia under a light microscope to identify the difference in sporulation of *F. zanthoxyli* on the three media (*Fusarium* can produce sickle-shaped large conidia and round small conidia). Calculate the concentration of spores using a hemocytometer.

2. Effects of Environmental Conditions on Spore Germination

Environmental conditions can affect the germination rate, speed of germination, length of germ tube, and even germination types of spores. In this experiment, *A. alternata* is used as the tested strain. This pathogen has a strong sporulation capacity and can ensure the number of spores required for the experiment.

①Preparation of spore suspension: Take the strain of *A. alternata* from the refrigerator, inoculate it on a PDA plate, and incubate it in the dark at 25℃ for 7 d. After that, 0.3% potato juice is added to the plate to wash the fungal spores off by wiping the surface of colony gently with a sterilized cotton swab, which is then filtered through a sterile double gauze to prepare spore suspensions. Take a drop of the spore suspensions and check it under a light microscope to adjust the spore concentration to approximately 1×10^6 spores/mL.

②Calculation of germination rate of spores: The standard for spore germination is that the length of the germ tube is greater than half the diameter of the spore. A total number of spores not less than 300 are measured. The calculation formula for germination rate is: germination rate (%) = number of germinated spores×100/total number of spores measured.

③Effects of temperature on spore germination: Use a pipette to draw 50 μL of spore suspensions dropwise into the groove of the concave glass slide, cover it with a coverslip, and place the slides in a 5℃, 10℃, 15℃, 20℃, 25℃, 30℃, 35℃ and 40℃ incubator, respectively. Every student can do two temperature treatments and each treatment contains three replicates. After 12 h of incubation, check the spore germination under a light microscope and calculate the spore germination rate at each temperature.

④Effects of humidity on spore germination: Take 50 μL of spore suspensions with a pipette into the groove of the concave glass slide, cover it with a coverslip, and place the slides in a sulfuric acid desiccator with relative humidity of 50%, 75%, 90% and 100%, respectively. The incubation temperature is 25℃. The sulfuric acid concentrations corresponding to the four humidity levels are 43.4%, 30.0%, 18.5% and 0, respectively. After 12 h of incubation, check the spore germination under a light microscope and calculate the spore germination rate under each humidity. Each treatment contains 3 replicates.

⑤Effects of pH on spore germination: Prepare sterile water with different pH values, 3, 4, 5, 6, 7, 8, 9, and 10, respectively. Take the spore suspensions and mix it with the same volume of sterile water at each pH value. Take 50 μL of the spore suspensions and drop them into the groove of the concave glass slide, cover it with a coverslip , and incubate it at 25℃ for 12 h.

Check the spore germination under a light microscope and calculate the spore germination rate under each humidity.

⑥Effects of nutrients on spore germination: Divide the spore suspensions into different centrifuge tubes in equal amounts, centrifuge them at 2 000 r/min for 3-5 min, discard the supernatant, add sterile water to the centrifuge tube to resuspend the spores debris at the bottom to wash the remaining medium. Repeat this procedure for 2-3 times. Add the same volume of 0.3% *P. orientalis* juice, 0.3% glucose solution, 0.3% sucrose solution and 0.3% potato juice to each centrifuge tube respectively, to completely suspended the spores in the nutrient solution. The spore suspension in sterile water serves as the control. Each treatment contains 3 replicates. After incubation at 25℃ for 12 h in the dark, the spore suspension in each centrifuge tube is used to prepare a temporary microscope slide. Check the spore germination under a light microscope and calculate the spore germination rate under each nutrient condition.

【Results and Discussion】

1. Calculate the spore concentration of *F. zanthoxyli* under different induction conditions, and analyze the influence of different conditions on the fungal sporulation.

2. Calculate the germination rate of *A. alternata* spores under different environmental conditions, and analyze the influence of environmental conditions on the spore germination.

3. Where can spore germination assay techniques be used in plant disease research?

4. What is the role of spores in plant pathogenesis and disease cycle?

实验二十二　受侵植物体内果胶酶活性的测定

【概述】

植物病原菌在与植物接触时，为了侵入到植物体内，要分泌一系列的酶类来降解植物的细胞壁，果胶酶是研究最多的细胞壁降解酶之一。植物病原菌通过分泌果胶酶降解植物细胞壁，使植物细胞彼此分离、组织软化，并且破坏植物细胞膜的渗透性，最终导致植物细胞的死亡，导致植物发病。果胶酶的活性越强，病原菌对植物细胞壁的破坏能力越大。

目前，主要采用DNS(3,5-二硝基水杨酸)方法测定果胶酶的活性。果胶酶降解果胶产生半乳糖醛酸，DNS与醛糖共热在碱性条件下产生棕红色氨基化合物，在一定范围内，还原糖含量与生成的棕红色物质的颜色成正比，在一定波长下，测定溶液的吸光值可算出半乳糖醛酸的含量，从而计算出果胶酶的活力。

【实验目的】

1. 了解果胶酶在植物病原物致病过程中的作用。

2. 学习并掌握植物体内果胶酶活性的测定方法和操作技术。

【材料和器具】

1. 实验材料

①供试植物：柑橘青霉病病果。

②实验试剂：D-半乳糖醛酸、果胶、果胶酶、DNS 试剂、四水合酒石酸钾钠、苯酚、无水亚硫酸钠、柠檬酸、磷酸氢二钠等。

2. 实验器具

水浴锅、离心机、96 微孔板、微孔板分光光度计、电子天平、试管、研钵、移液枪、枪头、离心管、记号笔等。

【方法和步骤】

1. 配制实验试剂

①DNS 试剂(3.15 g/L)：称取 3,5-二硝基水杨酸 7.5 g，NaOH 14.0 g，充分溶于 1 L 煮沸过的蒸馏水中，再加入四水合酒石酸钾钠 216 g 、苯酚 5.5 mL 和偏重亚硫酸钠 6.0 g，充分溶解后于棕色瓶中保存。

②D-半乳糖醛酸溶液(0.5 mg/mL)：精确称取 50 mg 半乳糖醛酸，用蒸馏水定容至 100 mL，获得 0.5 mg/mL 的 D-半乳糖醛酸溶液。

③果胶悬浮液(1 g/L)：称取 0.5 g 果胶，用蒸馏水加热搅拌溶解，用蒸馏水定容至 500 mL，放于 4℃冰箱保存，有效期 3 d。

④果胶酶溶液(5 mg/mL)：称取 500 mg 果胶酶，溶解于 100 mL 蒸馏水中，现用现配。

2. 绘制标准曲线

①取 8 支试管，编号，按表 2-4 所示加入各种试剂，混匀，在沸水浴中加热 5 min，迅速用流动水冷却，摇匀。

表 2-4 标准曲线绘制各试管反应物及用量

试剂体积(mL)	试管编号							
	1	2	3	4	5	6	7	8
半乳糖醛酸	0	0.2	0.4	0.6	0.8	1.0	1.2	1.4
蒸馏水	2.5	2.3	2.1	1.9	1.7	1.5	1.3	1.1
DNS	2.5	2.5	2.5	2.5	2.5	2.5	2.5	2.5

②以 1 号管为空白对照，测定各反应液在 540 nm 下的吸光值，以吸光值 A 为纵坐标，以反应体系中 D-半乳糖醛酸的量 x(mg)为横坐标，绘制标准曲线，并获得线性回归方程 $A=Kx+b$。

3. 样品果胶酶活性测定

①植物组织中果胶酶提取：取 1 g 柑橘青霉病病果果皮于研钵中，添加 5 mL 蒸馏水，研磨，离心，取上清液，即为果胶酶粗制品。

②果胶酶活力测定：取 12 支试管分别加入 2.0 mL 果胶悬浮液，在 50℃水浴中预热 10 min；取其中 6 支试管，其中 3 支加入植物组织提取果胶酶液 0.5 mL，另外 3 支加入煮沸 10 min 的植物组织提取果胶酶液作为对照；剩下 6 支试管，其中 3 支加入标准果胶酶液 0.5 mL，另外 3 支加入煮沸 10 min 的标准果胶酶液作为对照；立即混匀每支试管中的溶液，在 50℃水浴中反应 30 min；迅速加入 2.5 mL DNS 试剂，混匀后在沸水浴中显色反应 5 min，取出后迅速用流水冷却以终止显色反应；离心，取上清液于 96 微孔板中，用微孔

板分光光度计测定其在 540 nm 下的吸光值 A。

③计算酶活性：根据线性回归方程计算各处理的果胶酶活力(U)。

$$U=\frac{(A_1-A_2)-b}{K\times t\times M}$$

式中，A_1 为处理吸光值；A_2 为对照的吸光值；b 为标准曲线回归方程截距；K 为标准曲线斜率；t 为反应时间；M 为反应体系中酶量。

标准品果胶酶 M = 所取体积(mL)×标准品溶液浓度(mg/mL)

样品中果胶酶 M = 所取的粗酶液体积(mL)

【结果和讨论】

1. 计算所测果胶酶标准品和植物发病组织中果胶酶活性。
2. 谈谈果胶酶在病原物致病过程中的作用。
3. 可以用哪些方法测定果胶酶活性?

EXPERIMENT 22 Detection of Pectinase Activity in Infected Plant

【Introduction】

During the interaction between plants and pathogens, pathogens must secrete a series of enzymes to degrade plant cell walls to invade plants. Pectinase is one of the most well-studied cell wall degrading enzymes. Pathogens degrade plant cell walls by secreting pectinase to separate plant cells from each other, soften tissues, and damage the permeability of plant cell membranes, consequently leading to the death of plant cells and causing plant disease. The stronger activity of pectinase, the greater capacity of pathogens to damage plant cell walls.

At present, the DNS(3,5-dinitrosalicylic acid) assay is mainly used to determine the activity of pectinase. Pectinase degrades pectin to galacturonic acid. Brown-red amino compounds are produced when DNS and aldose are co-heated under alkaline conditions. Within a certain range, the content of reducing sugars is linear to the color of the brown-red substance. The content of galacturonic acid in the solution can be calculated by measuring the absorbance at a certain wavelength, and thus the activity of pectinase can be calculated.

【Experimental Purpose】

1. Understand the role of pectinase in the pathogenesis of plant pathogens.
2. Learn and master the methods and operations of pectinase activity measurement in plants.

【Materials and Apparatus】

1. Materials

①Test plant: *Citrus* fruit with *Penicillium* rot.

②Experimental reagents: D-galacturonic acid, pectin, pectinase, DNS reagent, potassium sodium tartrate tetrahydrate, phenol, anhydrous sodium sulfite, citric acid, disodium hydrogen phosphate, etc.

2. Instruments and Appliances

Water bath kettles, centrifuges, 96-well microplates, microplate spectrophotometers, electronic balance, test tubes, mortars, pipettes, pipette tips, centrifuge tubes, marking pens, etc.

【Methods and Procedures】

1. Preparation of Experimental Reagents

①DNS reagent(3.15 g/L): Weigh 7.5 g of 3,5-dinitrosalicylic acid and 14.0 g of sodium hydroxide, and dissolve them in 1 L of boiled distilled water. Add 216 g potassium tartrate tetrahydrate sodium, 5.5 mL phenol, and 6.0 g sodium metabisulfite to the solution. Fully dissolve and store it in a brown bottle for later use.

②D-galacturonic acid solution(0.5 mg/mL): Weigh 50 mg of galacturonic acid accurately and dilute it to 100 mL with distilled water to obtain a 0.5 mg/mL D-galacturonic acid solution.

③Pectin suspension(1 g/L): Add 0.5 g of pectin to distilled water, heat and stir it to dissolve, and dilute it to 500 mL with distilled water. Store the solution in a refrigerator at 4℃. The validity period is 3 d.

④Pectinase solution(5 mg/mL): Weigh 500 mg of pectinase and dissolve it in 100 mL of distilled water when it is used.

2. Make the Standard Curve

①Take 8 test tubes, number them, add various reagents as shown in Table 2-4, and mix them. Heat the tubes in a boiling water bath for 5 min, quickly cool them with running water, and shake them.

Table 2-4 Reactant and Dosage in Each Test Tube for Standard Curve

Reagent volume (mL)	Number of the test tube							
	1	2	3	4	5	6	7	8
D-galacturonic acid	0	0.2	0.4	0.6	0.8	1.0	1.2	1.4
Distilled water	2.5	2.3	2.1	1.9	1.7	1.5	1.3	1.1
DNS	2.5	2.5	2.5	2.5	2.5	2.5	2.5	2.5

②Take No. 1 tube as the blank control. Measure the absorbance of each reaction solution at 540 nm. Take absorbance A as Y-axis and the amount of D-galacturonic acid in the reaction system x(mg) as X-axis. Make a standard curve and obtain the linear regression equation $A=Kx+b$.

3. Determination of Pectinase Activity in Samples

①Extraction of pectinase from plant tissues: Take 1 g of the peel of citrus fruit with *Penicillium* rot and place the sample in a mortar, add 5 mL of distilled water, grind, centrifuge, and take the supernatant to obtain the crude pectinase extraction.

②Determination of pectinase activity: Take 12 test tubes and add 2.0 mL pectin suspension to each. Preheat the test tubes in a water bath at 50℃ for 10 min. Take 6 test tubes and add 0. 5 mL crude pectinase preparation to 3 of them. The other 3 are added with crude pectinase preparation which has been boiled for 10 min and serve as the control. Among the remaining 6 test tubes, 3 of which are added with 0. 5 mL standard pectinase solution, and the other 3 are added with standard pectinase solution which has been boiled for 10 minutes and serve as the control. Mix the solution in each test tube immediately and incubate them in a water bath at 50℃ for 30 min. Add 2. 5 mL DNS reagent quickly, mix, and incubate them for 5 min in a boiling water bath. Take them out and cool them quickly with running water to terminate the reaction. Centrifuge, take the supernatant into a 96-well plate, and measure the absorbance A at 540 nm with a microplate spectrophotometer.

③Calculation of enzyme activity: Calculate the pectinase activity (U) of each treatment according to the linear regression equation.

$$U=\frac{(A_1-A_2)-b}{K\times t\times M}$$

In the formula, A_1 is the absorbance value of the treatment; A_2 is the absorbance value of the control; b is the intercept of the standard curve regression equation; K is the slope of the standard curve; t is the reaction time; M is the amount of enzyme in the reaction system.

Standard pectinase M = volume (mL) × concentration of standard solution (mg/mL); pectinase M in the sample = volume of crude enzyme preparation(mL).

【Results and Discussion】

1. Calculate the pectinase activity in the tested pectinase standards and the plant disease tissues.
2. Talk about the role of pectinase in the pathogenesis of pathogens.
3. Are there any other methods can be used to determine pectinase activity?

实验二十三　受侵植物体内活性氧含量的测定

【概述】

植物受到病原物侵染后，其体内的活性氧(reactive oxygen species, ROS, 主要包括 O_2^- 和 H_2O_2)会被诱导产生，ROS 的产生与植物的抗病性密切相关。ROS 能直接杀伤植物病原菌，并通过参与植物细胞壁木质化及富含羟脯酸蛋白的氧化交联、寄主过敏性坏死反应和防卫基因表达信号的传递等提高寄主细胞的抵御入侵能力。

【实验目的】

1. 掌握植物体内活性氧测定方法和操作技术。
2. 了解活性氧在植物抗病性中的作用机制。

【材料和器具】

1. 实验材料

①供试病原：落叶松—杨栅锈菌(*Melampsora larici-populina*, MLP)的冷冻夏孢子。

②供试植物：1年生杨树，品种分别为：中林美荷、健杨和美洲黑杨，各品种8盆，均栽植于温室内。

③试剂：65 mmol/L 磷酸缓冲液(pH 7.8)，17 mmol/L 对氨基苯磺酸；7 mmol/L α-萘胺；10 mmol/L 盐酸羟胺；三氯甲烷萃取液、$NaNO_2$标准液、液氮等。

2. 实验器具

超低温冰箱、高速冷冻离心机、紫外分光光度计、恒温水浴锅、分析天平、移液器、研钵、剪刀、培养皿、离心管等。

【方法和步骤】

1. 杨树叶锈菌接种及采样

采用涂抹法将 MLP 的夏孢子接种在杨树叶片背面，具体操作如下：

①将保存于-80℃的 MLP 夏孢子粉倒入培养皿(直径 60 mm)中，称重并记录。

②将装有 MLP 夏孢子粉的培养皿吸湿活化 6~8 h，然后用干净的自来水稀释成孢子悬浮液，浓度为 1~2 mg/mL。

③用清水洗去杨树叶面尘土脏物，并用喷壶给叶背面喷雾，制造一层水膜。

④用毛笔蘸取 MLP 夏孢子悬浮液，均匀涂抹于杨树叶背面；以涂抹自来水为对照。在挂牌标签上做好标记，将其挂在杨树苗上。每个杨树品种接种 3~4 盆，每盆接种健康叶片 6 个，其中 3 个为 MLP 接种，3 个为对照。

⑤将接种杨树植株放置于保湿桶内避光保湿 24 h，对照植株单独放置；24 h 后将植株取出放置于温度为 22~26℃的温室中培养并观察。

⑥分别于接种后 24 h、48 h、72 h、120 h 和 168 h 观察各杨树品种发病情况，并采集各处理叶片 2.0 g 左右。每次采集叶片时，将叶片装入小塑料袋内后迅速放入装有液氮的保温杯内，带回实验室后，从液氮中取出材料放置在-80℃冰箱保存，备用。

2. 超氧阴离子自由基(O_2^-)含量的测定

本实验测定不同杨树品种接种 MLP 后细胞内超氧阴离子自由基的含量。O_2^- 含量测定采用羟胺氧化法。

O_2^- 能与羟胺溶液反应生成 NO_2^-：

$$NH_2OH+2O_2^-+H^+ \xlongequal{} NO_2^-+H_2O_2+H_2O$$

NO_2^- 经对氨基苯磺酸与 α-萘胺显色反应会生成红色产物对苯磺酸-偶氮-α-萘胺，该红色产物在 530 nm 有专一吸收峰。根据 NO_2^- 显色反应的标准曲线将 A_{530} 换算成 NO_2^- 浓度，在依据上述反应式直接进行 O_2^- 化学计量，即 NO_2^- 浓度乘以 2 得 O_2^- 浓度。

①$NaNO_2$标准液的配制：取预先在干燥器内放置 24 h 的 $NaNO_2$ 0.1 g，用蒸馏水完全溶解后，容量瓶定容至 1 000 mL，此溶液 NO_2^- 的浓度为 100 μg/mL，为 $NaNO_2$ 储备液，可储于棕色瓶中放冰箱保存。使用前，吸取 5 mL $NaNO_2$ 储备液，用蒸馏水定容至 100 mL，即为 $NaNO_2$ 标准液，浓度为 5 μg/mL。

②标准曲线的绘制：取 20 mL 的试管 7 支，编号后按表 2-5 添加试剂，每加一种试剂摇动试管使之混匀，加完后将试管放置在 30℃恒温水浴中保温 30 min，显色反应后测定反

应液在 530 nm 下的吸光值 A_{530}，以 NO_2^- 浓度为横坐标，A_{530}为纵坐标绘制标准曲线。

表 2-5　标准曲线绘制各试管反应物及用量

试剂体积(mL)	试管号						
	1(对照)	2	3	4	5	6	7
$NaNO_2$标准液	0	0.2	0.4	0.8	1.2	1.6	2.0
蒸馏水	2.0	1.8	1.6	1.2	0.8	0.4	0
对氨基苯磺酸	2.0	2.0	2.0	2.0	2.0	2.0	2.0
α-萘胺	2.0	2.0	2.0	2.0	2.0	2.0	2.0
每管 NO_2^- 含量(μg)	0	1.0	2.0	4.0	6.0	8.0	10

③O_2^- 的提取：分别取出待测杨树叶片组织样 0.5~1.0 g，迅速置于预冷的研钵中，液氮下充分研磨后转入 10 mL 离心管中，先加入 65 mmol/L(pH 7.8)磷酸缓冲液 5 mL，再用 5 mL 的磷酸缓冲液洗研钵并转入 10 mL 离心管，在 4℃下 12 000 r/min 离心 10 min，取上清液为 O_2^- 提取液。

④O_2^- 的测定：取 3 支 5 mL 的试管，分别加入 O_2^- 提取液 2 mL、65 mmoL/L(pH 7.8)磷酸缓冲液 1.5 mL、10 mmoL/L 盐酸羟胺 0.5 mL，混合后置于 25℃恒温水浴锅中水浴 20 min。另取 3 支试管，从上述 3 支试管各取反应液 2 mL 于新的 3 支试管中，向新的试管中分别加入 2 mL 对氨基苯磺酸溶液、2 mL α-萘胺，充分混匀后，放置于 30℃恒温水浴中反应 30 min，取 5 mL 反应液置于 10 mL 离心管中，加入 5 mL 三氯甲烷萃取，12 000 r/min 离心 3 min，取上层粉红色的水相测定 A_{530}。

⑤O_2^- 含量计算：从标准曲线上计算测定液对应的 NO_2^- 浓度，换算成 O_2^- 的浓度(X)后，可按以下公式计算出植物组织中的 O_2^- 含量：

$$O_2^- \text{含量}(\mu g/g) = (X \times Vt \times 2)/(W \times Vs)$$

式中，X 为测定液中 O_2^- 的浓度；Vt 为样品提取液总体积(mL)；2 为测定时样品提取液的稀释倍数；Vs 为显色反应时 O_2^- 提取液的量(mL)；W 为样品鲜重(g)。

【结果和讨论】

1. 计算各杨树品种在 MLP 侵染后细胞内的 O_2^- 含量，填入表 2-6，并结合症状观察，综合分析活性氧与杨树品种对 MLP 抗病性的关系。

表 2-6　受侵杨树品种中 O_2^- 含量的测定及反应型

杨树品种	处理及反应型	取样时间(h)				
		24	48	72	120	168
中林美荷	接种					
	对照					
	反应型					
健杨	接种					
	对照					
	反应型					

(续)

杨树品种	处理及反应型	取样时间(h)				
		24	48	72	120	168
美洲黑杨	接种					
	对照					
	反应型					

2. 查阅资料，讨论活性氧在植物体内的功能。

EXPERIMENT 23 Detection of Reactive Oxygen Species in Infected Plant

【Introduction】

The production of reactive oxygen species (ROS, mainly including O_2^- and H_2O_2) can be induced in plants infected by pathogens, which is closely related with plant disease resistance. ROS can directly kill plant pathogens and improve host resistance by involving in the lignification of plant cell walls, the oxidative cross-linking of hydroxyproline-rich proteins, host hypersensitive response, and the transduction of defense gene expression signals.

【Experimental Purpose】

1. Master the methods and operating techniques for measuring ROS in plants.
2. Understand the mechanism of ROS in plant disease resistance.

【Materials and Apparatus】

1. Materials

①Pathogen: Frozen uredospores of *Melampsora larici-populina* (MLP).

②Experimental plants: The varieties of one-year-old poplars, including Zhonglin Meihe (ZLMH), JianYang (JY) and Meizhou Heiyang (MZHY), each of which contains eight potting seedlings, incubated in the greenhouse.

③Reagents: 65 mmol/L phosphate buffer (pH 7.8), 17 mmol/L *p*-aminobenzene sulfonic acid, 7 mmol/L α-naphthylamine, 10 mmol/L hydroxylamine hydrochloride, chloroform extract, sodium nitrite standard solution, liquid nitrogen, etc.

2. Instruments and Appliances

Ultra-low temperature refrigerators, high-speed refrigerated centrifuges, UV spectrophotometers, constant temperature water baths, analytical balances, pipettes, mortar, scissors, Petri dishes, centrifuge tubes, etc.

【Methods and Procedures】

1. Inoculation of MLP and sampling

The uredospores of MLP are inoculatedon the back of poplar leaves using the smearing method. The specific operation is as follows:

①Pour the MLP uredospores stored at −80℃ into a Petri dish (diamater 60 mm), weigh it and record.

②Hygroscopically activate the MLP uredospores for 6-8 h in the Petri dish, and then dilute it with clean tap water to a spore suspension with a concentration of 1-2 mg/mL.

③Wash off the dust and dirt on the surface of poplar leaves with clean water. Spray on the back of the leaves to create a water film.

④Dip the MLP uredospore suspension with a writing brush, and evenly smear it on the back of the poplar leaves. Leaves treated with tap water serve as the control. Make a mark on the listing label and hang it on the poplar plant. Each poplar variety is inoculated with three to four plants, and each plant is inoculated with six healthy leaves, three for MLP inoculations and the other three for controls.

⑤Place the inoculated poplar plants in a moisturizing bucket for 24 h in the dark. The control plants are placed in a separate bucket. After 24 h, the plants are taken out and placed in a greenhouse at a temperature of 22-26℃ for incubation and observation.

⑥Observe the symptoms of each poplar variety at 24 h, 48 h, 72 h, 120 h and 168 h after inoculation. Collect about 2.0 g of leaves from each treatment. Put the sample into a small plastic bag and quickly put it into liquid nitrogen immediately. After bringing the samples to the laboratory, take the samples out of the liquid nitrogen and store them in a refrigerator at −80℃ for later use.

2. Content Determination of Superoxide Anion Radical

In this experiment, the content of superoxide anion radical (O_2^-) in the cells of different poplar varieties inoculated with MLP is determined by the hydroxylamine oxidation method.

O_2^- can react with hydroxylamine solution to generate NO_2^- according to the below chemical reaction equation:

$$NH_2OH + 2O_2^- + H^+ = NO_2^- + H_2O_2 + H_2O$$

NO_2^- reacts with *p*-aminobenzene sulfonic acid and α-naphthylamine to produce a red product, *p*-benzenesulfonic acid-azo-α-naphthylamine, which has a specific absorption peak at 530 nm. According to the standard curve of the color reaction of NO_2^-, A_{530} can be converted into NO_2^- concentration. Stoichiometry of O_2^- is directly carried out according to the above formula, that is, the O_2^- molarity is 2 folds of NO_2^- molarity.

①Preparation of $NaNO_2$ standard solution: Take 0.1 g of $NaNO_2$ that has been placed in a desiccator for 24 h in advance. Dissolve it in distilled water and adjust the volume to 1 000 mL. The NO_2^- concentration in this solution is 100 μg/mL. It is a $NaNO_2$ stock solution and can be stored in a brown bottle. Keep the bottle in the refrigerator. Before use, take 5 mL of $NaNO_2$ stock solution and dilute it to 100 mL with distilled water, which is the $NaNO_2$ standard solution at 5 μg/mL.

②Making the standard curve: Take seven 20-mL test tubes, add reagents according to Table 2-5 after numbering, and shake the test tube to mix reagents evenly. Incubate the test tubes in a water bath at 30℃ for 30 min. After the color reaction, measure the absorbance at 530 nm. Make a standard curve with NO_2^- concentration as X-axis and A_{530} as Y-axis.

③Extraction of O_2^-: Take 0.5-1.0 g of the poplar leaf tissue sample, place it in a pre-cooled mortar quickly, grind it thoroughly with liquid nitrogen and transfer it to a 10 mL centrifuge tube. Add 65 mmol/L(pH 7.8) phosphate buffer first, wash the mortar with 5 mL of phosphate buffer solution, and transfer it to the above centrifuge tube. The tube is then centrifuged at 12 000 r/min for 10 min at 4℃, and then the supernatant is taken as the O_2^- extract solution.

④Determination of O_2^-: Take three 5-mL test tubes, add 2 mL of O_2^- extract solution, 1.5 mL of phosphate buffer, and 0.5 mL of hydroxylamine hydrochloride. After homogenization, incubate the tubes in a water bath at 25℃ for 20 min. Transfer 2 mL of the reaction solution from each of the above test tubes to another 3 new test tubes, respectively. Add 2 mL of sulfanilic acid solution and 2 mL of α-naphthylamine to these test tubes. After homogenization, incubate the test tubes in a water bath at 30℃ for 30 min, take 5 mL of the reaction solution into a new 10-mL centrifuge tube, add 5 mL of chloroform for extraction, centrifuge at 12 000 r/min for 3 min, and take the upper pink water phase to determine A_{530}.

Table 2-5 Reactant and Dosage of Each Test Tube for Making of Standard Curve

Volume(mL)	No. of test tube						
	1(CK)	2	3	4	5	6	7
$NaNO_2$ standard solution	0	0.2	0.4	0.8	1.2	1.6	2.0
Distilled water	2.0	1.8	1.6	1.2	0.8	0.4	0
p-aminobenzene sulfonic acid	2.0	2.0	2.0	2.0	2.0	2.0	2.0
α-naphthylamine	2.0	2.0	2.0	2.0	2.0	2.0	2.0
NO_2^- content in each tube(μg)	0	1.0	2.0	4.0	6.0	8.0	10

⑤Calculation of O_2^- content: Calculate the NO_2^- concentration according to the standard curve and convert it to the O_2^- concentration(X), and then calculate the O_2^- content in the plant tissue according to the following formula:

$$X = NO_2^- \text{ content}(\mu g)/46 \times 32 \times 2$$

$$O_2^- \text{ content}(\mu g/g) = (X \times Vt \times 2)/(W \times Vs)$$

In the formula, X is the concentration of O_2^- of the tested solution; Vt is the total volume of the sample extract(mL); 2 is the dilution ratio of the sample extract during measurement; Vs is the amount of O_2^- extract during the color reaction(mL); W is the fresh weight of the sample(g).

【Results and Discussion】

1. Calculate the O_2^- content of each poplar variety after MLP infection, fill in Table 2-6, and combine the O_2^- content with an observation of symptoms to comprehensively analyze the relationship between ROS and the resistance of poplar varieties to MLP.

Table 2-6　Content of O_2^- and Reaction Type in Different Poplar Varieties

Poplar variety	Treatment andreaction type	Sampling time(h)				
		24	48	72	120	168
ZLMH	Inoculation					
	Control					
	Reaction type					
JY	Inoculation					
	Control					
	Reaction type					
MZHY	Inoculation					
	Control					
	Reaction type					

2. Please discuss the function of ROS in plants after searching and reading references.

实验二十四　受侵植物体内防御酶活性的测定

【概述】

植物病原菌在与植物接触时，寄主植物为了抵抗病原物的侵染与定植，在识别病原物后，会启动自身的防御反应，从而表现出不同程度的抗病性。与植物抗病性相关的重要指标之一为抗病相关酶。植物体内的抗病相关酶主要有苯丙氨酸解氨酶(PAL)、多酚氧化酶(PPO)、过氧化物酶(POD)、超氧化物歧化酶(SOD)、过氧化氢酶(CAT)等，这些酶活性越高，植物的抗病性相对越强。

【实验目的】

1. 了解抗病相关酶在植物抗病性中的作用。
2. 掌握植物体抗病相关酶活性测定方法和操作技能。

【材料和器具】

1. 实验材料

①供试病菌：花椒干腐病(*Fusarium zanthoxyli*)。

②供试植物：2 年生花椒盆栽苗，品种分别为凤县大红袍(FD)、韩城大红袍(HD)、豆椒(DJ)和野花椒(YHJ)，各品种 8~10 盆。

③实验试剂：L-苯丙氨酸、硼酸盐缓冲液(0.05 mol/L，pH 8.4，含有 5 mmol/L 的 β-巯基乙醇)、HCl 溶液(6 mol/L)、磷酸盐缓冲液(0.05 mol/L，pH 7.4)、磷酸盐缓冲液(0.05 mol/L，pH 8.8)、邻苯二酚水溶液(0.05 mol/L)、愈创木酚溶液(1%)、H_2O_2 溶液(1%，*V/V*)、PAL 酶联免疫分析试剂盒、PPO 酶联免疫分析试剂盒、POD 酶联免疫分析试剂盒、蒸馏水、PVPP、石英砂、液氮等。

2. 实验器具

天平、离心机、96 微孔板、微孔板分光光度计、恒温箱、研钵、单道移液器、8 通道移液器、剪枝剪、培养皿、离心管等。

【方法和步骤】

1. 花椒干腐病病原菌接种及采样

采用针刺接种法接种花椒干腐病病原菌。具体接种步骤如下:

①以在 PDA 培养基上培养 14 d 的干腐病病原菌为接种体,用无菌打孔器在菌落上制备菌饼,备用。

②采用无菌尖头镊子在花椒枝干上制造微伤口,用无菌牙签挑一个菌饼接种于针刺位点,然后用浸泡在无菌水中的脱脂棉片包裹接种位点,再采用保鲜膜包裹每个接种位点以进行保湿;以接种 PDA 培养基作为对照。每个品种接种 3 盆花椒苗,每个枝干上设 3 个接种位点,1 个对照,2 个接种。

③用挂牌标签做好标记,将其挂在花椒苗上。

④将接种好的花椒盆栽苗放置在温室内培养,分别于接种后 7 d(dpi)和 14 dpi 采用十字交叉法记录各花椒品种形成病斑的大小,并取样,保存在-80℃冰箱,备用。

⑤计算各花椒品种干腐病的病斑面积,病斑面积按照椭圆计算。

2. PAL 酶活性的测定

PAL 是植物体内苯丙烷代谢途径的关键酶,其控制着植物体内多种酚类、类黄酮植保素、木质素等抗菌物质的合成,因此 PAL 在植物的次生代谢和抗病代谢中具有重要作用。PAL 催化 L-苯丙氨酸脱氨生成反式肉桂酸,反式肉桂酸在 290 nm 处有强吸收峰,因此可通过测定反应液的 A_{290}值的变化计算 PAL 活性,主要操作步骤如下:

①PAL 粗酶液提取:称取 0.3 g 植物材料,剪碎,放置于研钵中,首先加入适量的液氮,迅速进行研磨,然后再向研钵中加入 3 mL 预冷的硼酸盐缓冲液(pH 8.4,含有 5 mmol/L 的β-巯基乙醇),再加入少许石英砂和 PVPP,继续研磨约 1 min,将研磨后的匀浆转移至 2 mL 离心管中,在 4℃ 15 000 r/min 离心 10 min 后取上清液,将上清液转移至 1.5 mL 离心管中,即为 PAL 粗酶液,备用。

②PAL 活性测定:采用微孔板分光光度计进行测定。96 微孔板每孔的反应体系为:150 μL 硼酸盐缓冲液、50 μL L-苯丙氨酸(0.02 mol/L)和 50 μL PAL 粗酶液,总反应体积为 250 μL,将微孔板放入微孔板分光光度计振荡摇匀,然后将微孔板置于 30℃ 恒温箱中反应 60 min,向每孔中加入 5 μL HCl(6 mol/L)终止反应,然后将微孔板置于微孔板分光光度计中在 290 nm 处测定吸光值,每个样品 3 次重复。空白对照:200 μL 硼酸盐缓冲液 和 50 μL PAL 粗酶液。

③PAL 活性计算:PAL 活性以每克鲜重组织每分钟使 A_{290}值增加 0.01 OD 为 1 个酶活单位(U),计算公式如下:

$$\text{PAL 活性}[\text{U}/(\text{g}\cdot\text{min})]=\frac{A_{290}\times Vt\times v}{0.01\times Vs\times M\times t}$$

式中,A_{290}为酶促反应的 OD 值;Vt 为 PAL 提取液总体积(3 000 μL);Vs 为测定时取酶液的体积(50 μL)；v 为反应液总体积(250 μL)；M 为样品鲜重(0.3 g)；t 为酶促反应时间(60 min)。

3. PPO 酶活性的测定

PPO 广泛存在于植物体内，能催化多酚类物质氧化成醌类，也能直接以 O_2 为底物将酚氧化成醌类，此外，PPO 也参与木质素的形成，从而抑制病原物的侵染。PPO 催化邻苯二酚形成褐色的醌，在分光光度计 410 nm 处使反应体系的 OD 值产生变化，通过 OD 值上升的读数变化确定 PPO 的酶活大小。实验步骤如下：

①PPO 粗酶液提取：称取 0.3 g 植物材料，剪碎，放置于研钵中，加入适量的液氮，迅速进行研磨，向液氮研磨后的研钵中加入 3 mL 预冷的磷酸盐缓冲液(pH 7.4)，再加入少许石英砂和 PVPP，继续研磨约 1 min，将研磨后的匀浆转移至 2 mL 离心管中，在 4℃、15 000 r/min 离心 10 min 后取上清液，将上清液转移至 1.5 mL 离心管中备用。

②PPO 活性测定：采用微孔板分光光度计进行测定。首先向 96 微孔板的每孔添加 50 μL PBS(0.05 mol/L，pH 8.8)和 100 μL 邻苯二酚(0.05 mol/L)，混匀后在 30℃恒温箱中孵育 5 min，然后每孔中添加 50 μL PPO 粗酶液，混匀后静置 5 min，然后读取反应液在 410 nm 处的吸光值。每个样品 3 次重复。空白对照：150 μL PBS 和 50 μL PAL 粗酶液。

③PPO 活性计算：PPO 活性以每克鲜重组织每分钟使 A_{410} 值增加 0.01 OD 为 1 个酶活单位(U)，计算公式如下：

$$\text{PPO 活性}[\text{U/(g}\cdot\text{min)}]=\frac{A_{410}\times Vt\times v}{0.01\times Vs\times M\times t}$$

式中，A_{410} 为酶促反应的 OD 值；Vt 为 PPO 提取液总体积(3 000 μL)；Vs 为测定时取酶液的体积(50 μL)；v 为反应液总体积(200 μL)；M 为样品鲜重(0.3 g)；t 为酶促反应时间(5 min)。

4. POD 酶活性的测定

POD 能清除植物体内的活性氧自由基，可提高植物的抗逆性。在有 H_2O_2 存在的条件下，POD 可催化愈创木酚氧化形成茶褐色的 4-邻甲氧基苯酚，该物质的最大吸收波长为 470 nm。可用分光光度计测定 470 nm 的吸光值变化反映 POD 的活性。实验步骤如下：

①POD 粗酶液提取：POD 粗酶液提取同 PPO。

②POD 活性测定：采用微孔板分光光度计进行测定。96 微孔板每孔的反应体系为：150 μL PBS(0.05 mol/L，pH 8.8)、25 μL 愈创木酚(1%)、25 μL H_2O_2(1%)和 10 μL POD 粗酶液，混匀后在 37℃恒温箱中反应 10 min，测定其在 470 nm 处的吸光值，每个样品 3 次重复。空白对照为将 25 μL H_2O_2 更换为 25 μL PBS，其他组分不变。

③POD 活性计算：POD 活性以每克鲜重组织每分钟使 A_{470} 值增加 0.01 OD 为 1 个酶活单位(U)，计算公式如下：

$$\text{POD 活性}[\text{U/(g}\cdot\text{min)}]=\frac{A_{470}\times Vt\times v}{0.01\times Vs\times M\times t}$$

式中，A_{470}为酶促反应的 OD 值；Vt 为 PPO 提取液总体积(3 000 μL)；Vs 为测定时取酶液的体积(10 μL)；v 为反应液总体积(210 μL)；M 为样品鲜重(0.3 g)；t 为酶促反应时间(10 min)。

5. 酶联免疫分析试剂盒检测酶活性

为了实现样品的高通量检测，可以采用酶联免疫分析试剂盒对各种酶活性进行检测。其原理是将纯化的植物 PAL(POD 或 PPO)的抗体包被在微孔板上，制成固相抗体，向包被单抗的微孔中加入 PAL(POD 或 PPO)提取液，再与辣根过氧化物酶(HRP)标记的 PAL(POD 或 PPO)抗体结合，形成抗体—抗原—酶标抗体复合物，经过彻底洗涤后加底物 TMB(3,3′,5,5′-四甲基联苯胺)显色。TMB 在 HRP 酶的催化下转化成蓝色，并在酸的作用下最终转化成黄色。颜色的深浅和样品中的 PAL(POD 或 PPO)呈正相关。用酶标仪在 450 nm 波长下测定吸光度，通过标准曲线计算样品中植物 PAL(POD 或 PPO)浓度。

具体操作过程可依据试剂盒说明书进行。

【结果和讨论】

1. 详细记录干腐病在供试花椒品种枝干上发病情况，填入表 2-7，分析各花椒品种的抗病性差异。

表 2-7 各花椒品种枝干上的病斑面积

花椒品种	病斑面积(mm^2)	
	7 dpi	14 dpi
FD		
HD		
DJ		
YHJ		

2. 计算各花椒品种受侵后体内抗病相关酶活性，填写表 2-8，并利用实验数据绘图；关联分析抗病相关酶活性与不同花椒品种对干腐病的抗病性差异之间的联系。

表 2-8 各受侵花椒品种枝干中抗病相关酶活性测定值 [U/(g · min)]

花椒品种	处理	PAL		PPO		POD	
		7 dpi	14 dpi	7 dpi	14 dpi	7 dpi	14 dpi
FD	接种						
	对照						
HD	接种						
	对照						
DJ	接种						
	对照						
YHJ	接种						
	对照						

3. 植物的抗病机制有哪些类型？
4. 谈谈如何利用植物的抗病性。

EXPERIMENT 24 Detection of Activities of Defensive Enzymes in Infected Plant

【Introduction】

When plant pathogens come into contact with plants, in order to resist the infection and colonization of pathogens, the host plants will initiate their own defense responses after recognizing the pathogens, thus showing different levels of disease resistance. One of the important indicators related to plant disease resistance is disease resistance-related enzymes. Disease resistance-related enzymes in plants mainly include phenylalanine ammonia lyase (PAL), polyphenol oxidase (PPO), peroxidase (POD), superoxide dismutase (SOD), catalase (CAT), etc. The higher activity of these enzymes, the stronger disease resistance of plants.

【Experimental Purpose】

1. Understand the role of disease resistance-related enzymes in plant disease resistance.

2. Master the method and operation skills of determining the activity of plant disease resistance-related enzyme.

【Materials and Apparatus】

1. Materials

①Pathogen: The pathogen of *Zanthoxylum bungeanum* stem canker (*Fusarium zanthoxyli*).

②Tested plants: Two-year-old potted seedlings of *Z. bungeanum*. The varieties are Fengxian Dahongpao (FD), Hancheng Dahongpao (HD), Doujiao (DJ), and wild prickly ash (YHJ), with eight to ten seedlings of each variety.

③Reagents: L-phenylalanine, borate buffer (0.05 mol/L, pH 8.4, containing 5 mmol/L β-mercaptoethanol), HCl solution (6 mol/L), phosphate buffer (0.05 mol/L, pH 7.4), phosphate buffer (0.05 mol/L, pH 8.8), catechol aqueous solution (0.05 mol/L), guaiacol solution (1%), H_2O_2 solution (1%), PAL ELISA kit, PPO ELISA kit, POD ELISA kit, distilled water, PVPP, quartz sands, liquid nitrogen, etc.

2. Instruments and Appliances

Electronic balance, centrifuges, 96-well microplates, microplate spectrophotometers, incubators, mortars, single-channel pipettes, 8-channel pipettes, pruning shears, Petri dishes, centrifuge tubes, etc.

【Methods and Procedures】

1. Inoculation of *F. zanthoxyli* and Sampling

The inoculation of *F. zanthoxyli* is conducted using the method of acupuncture inoculation. The specific inoculation steps are as follows:

①The pathogen *F. zanthoxyli* cultured on a PDA medium for 14 d is taken as the inoculum. Prepare fungal discs from the colony with a sterile puncher.

②Make micro wounds on the branches of *Z. bungeanum* with sterile pointed tweezers. Pick a fungal disc with a sterile toothpick and inoculate it to the acupuncture site. Wrap the inoculation site with a piece of water-soaked absorbent cotton, and then wrap each inoculation site with cling film for moisture retention. The inoculation of the PDA medium serves as the control. Each variety is inoculated with three *Z. bungeanum* seedlings, with three inoculation sites on each branch, one for control, and the other two for inoculations.

③Mark the seedlings with a listing label.

④The inoculated *Z. bungeanum* seedlings are incubated in the greenhouse. Record the size of the lesions on each *Z. bungeanum* variety with the cross method at 7 dpi and 14 dpi, respectively. Collect samples and store them at −80℃ for later use.

⑤Calculate the lesion area on each *Z. bungeanum* variety. The lesion area is calculated as an ellipse.

2. Determination of PAL Enzyme Activity

PAL is a key enzyme in the metabolic pathway of phenylpropane in plants. It controls the synthesis of various phenols, flavonoid phytoalexins, lignins and other antimicrobial substances in plants. Therefore, PAL plays an important role in the secondary metabolism and disease-resistant metabolism of plants. PAL catalyzes the deamination of L-phenylalanine to trans-cinnamic acid, which has a strong absorption peak at 290 nm. Therefore, the PAL activity can be calculated by measuring the change of A_{290} in the reaction solution. The main operating procedures are as follows:

①Preparation of PAL crude enzyme solution: Weigh 0. 3 g of plant material, cut it into small pieces, and quickly grind it in a mortar with liquid nitrogen. Add 3 mL of pre-cooled borate buffer(pH 8. 4, containing 5 mmol/L β-mercaptoethanol), add a little quartz sand and PVPP, and homogenate it for about 1 min. Transfer the mixture to a 2 mL centrifuge tube, centrifuge at 4℃, 15 000 r/min for 10 min, and then transfer the supernatant to a 1. 5-mL centrifuge tube. This supernatant is the crude PAL enzyme solution.

②PAL activity determination: Measure the PAL activity with a microplate branch photometer. The reaction mixtures in each well of 96-well plate contain 150 μL borate buffer, 50 μL L-phenylalanine(0. 02 mol/L) and 50 μL PAL crude enzyme solution, and the total reaction volume is 250 μL. Put the 96-well plate into the microplate spectrometer and shake it evenly, then place the 96-well plate in a 30℃ incubator for 60 min. Add 5 μL HCl(6 mol/L) to each well to stop the reaction, place the 96-well plate in the spectrometer, and measure the absorbance at 290 nm, and each sample is repeated three times. The mixtures of blank control in each well contain 200 μL borate buffer and 50 μL PAL crude enzyme solution.

③PAL activity calculation: PAL activity increases the A_{290} value by 0. 01 OD per gram of fresh tissue per minute as an enzyme activity unit(U). The calculation formula is as follows:

$$\text{PAL activity}[\text{U}/(\text{g}\cdot\text{min})]=\frac{A_{290}\times Vt\times v}{0.01\times Vs\times M\times t}$$

In the formula, A_{290} is the OD value of the enzymatic reaction; *Vt* is the total volume of the PAL extract(3 000 μL); *Vs* is the volume of the enzyme solution in each reaction(50 μL); *v* is the total volume of the reaction solution(250 μL); *M* is the fresh weight of the sample(0. 3 g); *t* is the enzymatic reaction time(60 min).

3. Determination of PPO Enzyme Activity

PPO is widely present in plants. It can catalyze the oxidation of polyphenols into quinones, and can also directly oxidize phenols into quinones with O_2 as a substrate. In addition, PPO also participates in the formation of lignin, thereby inhibits the infection of pathogens. PPO catalyzes the formation of brown quinone from catechol, and changes the OD value of the reaction solution at 410 nm. The enzyme activity of PPO is determined by the change of the OD value. The experimental procedures are as follows:

①Extraction of PPO crude enzyme solution: Weigh 0. 3 g of plant material, cut it into small pieces, and quickly grind it in liquid nitrogen. Add 3 mL of pre-cooled phosphate buffer(pH 7. 4) and a little quartz sand and PVPP to the mortar after grinding. Homogenate it for about 1 minute and transfer the mixture to a 2 mL centrifuge tube. Centrifuge at 4℃, 15 000 r/min for 10 minutes and transfer the supernatant to 1. 5 mL centrifuge tube for later use.

②Determination of PPO activity: Measure with a microplate photometer. First, add 50 μL PBS(0. 05 mol/L, pH 8. 8) and 100 μL catechol(0. 05 mol/L) to each well of a 96-well plate, mix, and then incubate in a 30℃ incubator for 5 min. Add 50 μL of crude PPO enzyme solution to each well, mix and let it stand for 5 min. Measure the absorbance value of the reaction solution at 410 nm. Each reaction contains 3 technical replicates. The mixtures of blank control in each well contain 150 μL PBS and 50 μL PAL crude enzyme solution.

③Calculation of PPO activity: The PPO activity is expressed as an enzyme activity unit (U), where one unit represents the increase in the rate of 0. 01 OD per gram of fresh tissue per minute. The calculation formula is as follows:

$$\text{PPO activity}[\text{U}/(\text{g}\cdot\text{min})]=\frac{A_{410}\times Vt\times v}{0.01\times Vs\times M\times t}$$

In the formula, A_{410} is the OD value of the enzymatic reaction; *Vt* is the total volume of the PAL extract(3 000 μL); *Vs* is the volume of the enzyme solution in each reaction(50 μL); *v* is the total volume of the reaction solution(250 μL); *M* is the fresh weight of the sample(0. 3 g); *t* is the enzymatic reaction time(60 min).

4. Determination of POD Enzyme Activity

POD can scavenge active oxygen free radicals in plants and improve the resistance of plants. In the presence of H_2O_2, POD can catalyze the oxidation of guaiacol to 4-O-methoxyphenol with dark brown colour, which has a maximum absorption at 470 nm. The change in absorbance at 470 nm can be measured with a spectrophotometer to reflect the activity of POD. The experimental procedures are as follows:

①Extraction of POD crude enzyme solution: The extraction of crude POD enzyme solution is the same as that of PPO.

② POD activity determination: Measure with a microplate photometer. The reaction mixtures in each well of the 96-well plate contain 150 μL PBS(0.05 mol/L, pH 8.8), 25 μL guaiacol(1%), 25 μL H_2O_2(1%, V/V) and 10 μL POD. After mixing, place it in a 37℃ incubator for 10 min, and measure the absorbance at 470 nm. Each reaction contains 3 technical replicates. In the blank control, 25 μL PBS replaces 25 μL H_2O_2, with the other components remain unchanged.

③Calculation of POD activity: The PPO activity is expressed as an enzyme activity unit (U), where one unit represents the increase in the rate of 0.01 OD per gram of fresh tissue per minute. The calculation formula is as follows:

$$\text{PPO activity}[\text{U}/(\text{g} \cdot \text{min})] = \frac{A_{470} \times Vt \times v}{0.01 \times Vs \times M \times t}$$

In the formula, A_{470} is the OD value of the enzymatic reaction; Vt is the total volume of the PPO extract(3 000 μL); Vs is the volume of the enzyme solution in each reaction(10 μL); v is the total volume of the reaction solution(210 μL); M is the fresh weight of the sample(0.3 g); t is the enzymatic reaction time(10 min).

5. Determination of Enzyme Activity Using ELISA Kits

In order to achieve high-throughput detection of samples, ELISA kits can be used to detect various enzyme activities. The principle is to make a solid phase antibody by coating the purified plant PAL(POD or PPO) antibody on a microtiter plate. Add the PAL(POD or PPO) extract to the microwell of the coated monoclonal antibody, and then mix it with horseradish root peroxidase(HRP) labeled PAL(POD or PPO) antibody to form an antibody-antigen-enzyme labeled antibody complex. After thorough washing, the substrate TMB(3,3′,5,5′-tetramethylbenzidine) is added for a coloring reaction. TMB is converted to blue under the catalysis of HRP enzyme and finally changed to yellow under acid. The intensity of the color is positively related to the PAL(POD or PPO) in the sample. Measurethe absorbance with a microplate reader at 450 nm and calculate the concentration of plant PAL (POD or PPO) in the sample according to the standard curve.

The specific operation process can be carried out according to the instructions of the kit.

【Results and Discussion】

1. Record the occurrence of dry rot on the branches and stems of the tested *Z. bungeanum* varieties in detail, and fill in Table 2-7 to analyze the difference in disease resistance of each *Z. bungeanum* variety.

2. Calculate the activities of defensive enzymes of each *Z. bungeanum* variety after infection, fill in Table 2-8, and plot the experimental data. Analyze the relationship between activities of defensive enzymes and the resistance of different *Z. bungeanum* varieties to stem canker.

Table 2-7　Lesion Areas on Branches of *Z. bungeanum* Varieties

Tested plants	Lesion area(mm²)	
	7 dpi	14 dpi
FD		
HD		
DJ		
YHJ		

Table 2-8　Activities of Defensive Enzymes in the Stems of *Z. bungeanum*

[U/(g · min)]

Tested plants	Treatment	PAL		PPO		POD	
		7 dpi	14 dpi	7 dpi	14 dpi	7 dpi	14 dpi
FD	Inoculation						
	Control						
HD	Inoculation						
	Control						
DJ	Inoculation						
	Control						
YHJ	Inoculation						
	Control						

3. What are the types of plant disease resistance mechanisms?
4. How to use the disease resistance of plants?

实验二十五　杀菌剂的室内毒力测定

【概述】

化学防治是植物病害防治的主要措施之一。筛选高效、无公害杀菌剂，了解杀菌剂对植物病原菌的杀菌能力及其作用范围是实施田间药剂防治的前提。杀菌剂毒力测定常用的方法有孢子萌发法、生长速率法和抑菌圈法等。孢子萌发法是用不同浓度供试药剂处理病原真菌孢子，在合适温度下培养一定时间后检查孢子的萌发率。生长速率法适用于营养菌丝生长快、且不易产生孢子的病原真菌杀菌剂毒力测定，常用的方法是将不同药剂分别放在人工培养基平板中，再接种病菌后比较不同平板上的菌落大小来确定药剂抑菌程度。

【实验目的】

1. 掌握杀菌剂室内毒力测定方法与操作技术。
2. 学习杀菌剂药效评价方法。

【材料和器具】

1. 实验材料

①供试病菌：花椒根腐病(*Fusarium solani*)。

②供试杀菌剂：80%代森锰锌可湿性粉剂、80%嘧菌酯水分散粒剂、47%春雷·王铜可湿性粉剂和50%多菌灵可湿性粉剂。

③其他：PDA培养基、PD液体培养基、无菌水等。

2. 实验器具

灭菌培养皿、三角瓶、小烧杯、离心管、移液器、枪头、载玻片、盖玻片、打孔器、玻璃棒、牙签、记号笔、尖嘴镊子、酒精灯、橡皮筋等。

【方法和步骤】

1. 孢子萌发法

①孢子悬浮液的制备：从冰箱中取出花椒根腐病病原菌菌种，接种于PDA培养基平板上，25℃暗培养7 d后，添加0.3% PD液体培养基，用灭菌的棉签在培养基表面轻轻擦拭，洗下真菌孢子，用灭菌的双层纱布过滤，制备成孢子悬浮液。取一滴孢子悬浮液，显微镜下检查，调整孢子浓度为1×10^6个/mL左右，备用。

②杀菌剂药液的配制：用无菌水配置供试杀菌剂的母液(1 mg/mL)，然后用无菌水分别配置浓度为0.05 mg/mL、0.1 mg/mL、0.2 mg/mL、0.3 mg/mL、0.4 mg/mL和0.5 mg/mL的药液。

③孢子与药液共培养：取无菌离心管，向管内添加500 μL孢子悬浮液和500 μL药液，每个浓度3个重复，离心管盖好盖子后，将其放置于25℃恒温振荡培养箱内培养(150 r/min)。以500 μL无菌水代替药液作为对照。

④孢子萌发检测：分别在药剂处理后12 h、24 h、36 h和48 h，吸取少许孢子悬浮液，制作临时显微玻片，在显微镜下检查孢子萌发情况，计算各药剂处理下的孢子萌发率(各处理检查孢子总数不低于300个)，获得各药剂处理对孢子萌发的抑制率。

$$\text{萌发率}(\%)=\frac{N\times100}{Nt-5}$$

式中，N为萌发的孢子数目；Nt为检测的孢子总数。

$$\text{孢子萌发抑制率}(\%)=\frac{(G_0-G)\times100}{G_0}$$

式中，G_0为对照的孢子萌发率；G为杀菌剂处理的孢子萌发率。

⑤药效评价：有效中浓度(median effective concentration，EC_{50})能反应药剂对病原物的抑菌作用。EC_{50}越小，说明抑菌能力越强，反之亦然。将抑菌率转换成概率值，获得概率值与药剂浓度之间的线性回归方程，计算出各药剂的EC_{50}值。

2. 菌落生长速率法

①病原菌的活化培养：从冰箱中取出花椒根腐病病原菌菌种，接种于PDA培养基平板上，25℃暗培养7 d后，备用。

②杀菌剂药液的配制：用无菌水分别配置供试杀菌剂的母液(1 mg/mL)，然后用无菌水分别配置浓度为0.05 mg/mL、0.1 mg/mL、0.2 mg/mL、0.3 mg/mL、0.4 mg/mL和0.5 mg/mL的药液。

③含药培养基平板的配制：加热融化提前配置好、灭菌过的PDA培养基，取49 mL

倒入 100 mL 无菌三角瓶中，冷却至 50~55℃时分别加入 1 mL 不同浓度供试药剂，对照加入 1 mL 无菌水，充分摇匀后迅速倒入无菌培养皿中，各处理浓度 3 次重复。

④病原菌的接种：用无菌打孔器在培养好的花椒根腐病菌菌落上制备菌饼(直径 5 mm)，用无菌牙签挑取 1 个菌饼接种至各处理平板培养基中央，置于 25℃恒温箱内培养 3~7 d。

⑤菌落生长抑制率的计算：培养 3~7 d 后观察并测量各处理菌落直径(mm)。每个菌落十字交叉测量两次，取其平均值为其菌落直径，计算各处理对病菌菌落生长的抑制率。

$$\text{菌落生长抑制率}(\%)=\frac{(R_0-R)\times 100}{R_0-5}$$

式中，R_0 为对照的菌落直径；R 为杀菌剂处理的菌落直径。

⑥药效评价：将抑菌率转换成概率值，获得概率值与药剂浓度之间的线性回归方程，计算出各药剂对菌落生长抑制作用的 EC_{50} 值。

【结果和讨论】

1. 详细记录孢子萌发法评价各供试药剂对花椒根腐病病原菌的抑菌作用数据，填写表 2-9，比较分析各药剂的药效。

表 2-9　各供试药剂对孢子萌发抑制效果评价

供试杀菌剂	线性回归方程	相关系数	EC_{50}
80%代森锰锌可湿性粉剂			
80%嘧菌酯水分散粒剂			
47%春雷・王铜可湿性粉剂			
50%多菌灵可湿性粉剂			

2. 详细记录生长速率法评价各供试药剂对花椒根腐病菌的抑菌作用数据，填写表 2-10，分析比较各药剂的药效。

表 2-10　各供试药剂对菌落生长抑制效果评价

供试杀菌剂	线性回归方程	相关系数	EC_{50}
80%代森锰锌可湿性粉剂			
80%嘧菌酯水分散粒剂			
47%春雷・王铜可湿性粉剂			
50%多菌灵可湿性粉剂			

3. 综合孢子萌发和生长速率法实验数据，综合评价各供试杀菌剂对花椒根腐病病原菌的抑菌活性。

4. 谈谈进行杀菌剂室内毒力测定的必要性。

EXPERIMENT 25　Bioassay of Toxicity of Pesticide *in vitro*

【Introduction】

Chemical control is one of the main strategies for plant disease management. The screening of high-efficiency and non-pollution fungicides is a prerequisite for the implementation of field chemical control, so as to understand the sterilization ability of a certain fungicide against plant pathogens and its scope of action. The common methods for determining the virulence of fungicides include spore germination method, growth rate method and inhibition zone method. The spore germination method is to treat the pathogenic fungal spores with different concentrations of test fungicides, and then check the germination rate of the spores after incubating for a certain time at a suitable temperature. The growth rate method is suitable for the determination of the virulence of pathogenic fungi that have fast growth of vegetative hyphae and are not easy to produce spores. The common method is to add different fungicides to artificial medium plates, and then compare the sizes of colonies on different plates after inoculation with pathogens to determine the antifungal effect of the fungicide.

【Experimental Purpose】

1. Master the detection methods and operation techniques of the toxicity tests of fungicides *in vitro*.

2. Learn how to evaluate the efficacy of fungicides.

【Marerials and Apparatus】

1. Materials

①Pathogen: the pathogen of *Zanthoxylum bungenum* root rot(*Fusarium solani*).

②Tested pesticides: 80% mancozeb wettable powder (WP), 80% azoxystrobin water dispersible granules(WDG), 47% Kasumin-Bordeaux(WP), and 50% carbendazim(WP).

③Others: PDA medium, PD liquid medium, sterile water, etc.

2. Instruments and Appliances

Sterilized Petri dishes, Erlenmeyer flasks, small beakers, centrifuge tubes, pipettes, pipette tips, glass slides, coverslips, hole punches, glass rods, toothpicks, markers, pointed tweezers, alcohol lamps, rubber bands, etc.

【Methods and Procedures】

1. Spore Germination Method

①Preparation of spore suspension: Take *F. solani* from the refrigerator, inoculate it on a PDA plate, and culture in the dark at 25 ℃ for 7 d. Add 0.3% PD medium, wipe the surface of the culture gently with a sterilized cotton swab to wash off the fungal spores, and filter with two layers of sterilized cheese cloth to prepare a spore suspension. Take a drop of the spore suspension and check it under a light microscope to adjust the spore concentration to about 1×10^6/mL for later use.

②Preparation of pesticide solution: Prepare the stocking solution (1 mg/mL) of the test pesticide with sterile water, and then use sterile water to prepare the solutions at 0.05 mg/mL, 0.1 mg/mL, 0.2 mg/mL, 0.3 mg/mL, 0.4 mg/mL, and 0.5 mg/mL.

③Co-cultivation of spores and pesticide solution: Add 500 μL of spore suspension and 500 μL of pesticide solution to a sterile centrifuge tube, with 3 replicates for each concentration. Cap the centrifuge tube and place it at 25℃ in a constant temperature shaking incubator (150 r/min). Use 500 μL of sterile water instead of the pesticide solution as the control.

④Detection of spore germination: Take a little spore suspension and make temporary slides to check the spore germination under a light microscope respectively at 12 h, 24 h, 36 h and 48 h after the pesticide treatment. Calculate the spore germination rate under the treatment of each pesticide (check no less than 300 spores in each treatment), and obtain the inhibitory rate of the spore germination in each fungicide treatment.

$$\text{Germination rate}(\%) = \frac{N \times 100}{Nt-5}$$

In the formula, N is the number of germinated spores; Nt is the total spores inspected.

$$\text{Spore germination inhibitory rate}(\%) = \frac{(G_0-G) \times 100}{G_0}$$

In the formula, G_0 is the germination rate of control; G_0 is the germination rate of pesticide treatment.

⑤Efficacy evaluation of pesticide: Median effective concentration (EC_{50}) can reflect the antifungal effect of the pesticide on pathogens. The smaller the EC_{50}, the stronger the antifungal ability, and vice versa. Convert the inhibitory rate to a probability value, obtain the linear regression equation between the probability value and the concentration of the fungicide, and calculate the EC_{50} value of each pesticide.

2. Colony Growth Rate Method

①Activation culture of pathogen: Take *F. solani* from the refrigerator, inoculate it on a PDA plate and culture it in the dark at 25℃ for 7 d.

②Preparation of pesticide solution: Use sterile water to prepare the stocking solution (1 mg/mL) of the tested pesticide, and then use sterile water to prepare the pesticide solution at the concentration of 0.05 mg/mL, 0.1 mg/mL, 0.2 mg/mL, 0.3 mg/mL, 0.4 mg/mL and 0.5 mg/mL.

③Preparation of the pesticide plate: Heat and melt the pre-prepared and sterilized PDA medium, pour 49 mL PDA into a 100 mL sterile Erlenmeyer flask, add 1 mL of different concentrations of pesticide solution to PDA when cooled to 50−55℃, respectively. Add 1 mL of sterile water instead of pesticide solution to the control. Shake well and then quickly pour it into a sterile Petri dish. Each treatment concentration contains 3 replicates.

④Inoculation of pathogen: Prepare a disc (5 mm in diameter) on *F. solani* colony with a sterile puncher. Pick a disc with a sterile toothpick and inoculate it on the center of each plate. Incubate it at 25℃ for three to seven days.

⑤Calculation of colony inhibition rate: Observe and measure the colony diameter (mm) of each treatment after 3-7 d of incubation. Each colony is cross-measured twice, and the average value is taken as its colony diameter. Calculate the inhibition rate of each treatment on the growth of the fungal colony.

$$\text{Colony inhibitory rate}(\%) = \frac{(R_0 - R) \times 100}{R_0 - 5}$$

In the formula, R_0 is the diameter of the control; R is the diameter of treatment.

⑥Efficacy evaluation of pesticide: Convert the inhibitory rate into a probability value, obtain a linear regression equation between the probability value and the concentration of the pesticide, and calculate the EC_{50} value of the inhibitory effect of each fungicide on the growth of colony.

【Results and Discussion】

1. Record the antifungal data of each tested fungicide on *F. solani* using the spore germination method in detail and fill in Table 2-9. Analyze and compare the efficacy of each fungicide.

Table 2-9 Evaluation of Inhibitory Effect of Each Tested Pesticide on Spore Germination

Pesticides	Linear regression equation	Correlation coefficient	EC_{50}
80% Mancozeb(WP)			
80% Azoxystrobin(WDG)			
47% Kasumin-Bordeaux(WP)			
50% Carbendazim(WP)			

2. Record the antibacterial data of each tested fungicide on *F. solani* using growth rate method in detail, and fill in Table 2-10. Analyze and compare the efficacy of each fungicide.

Table 2-10 Evaluation of Inhibitory Effect of Each Tested Pesticide on Colony Growth

Pesticides	Linear regression equation	Correlation coefficient	EC_{50}
80% Mancozeb(WP)			
80% Azoxystrobin(WDG)			
47% Kasumin-Bordeaux(WP)			
50% Carbendazim(WP)			

3. Combine the experimental data of spore germination method and growth rate method, and comprehensively evaluate the antifungal effect of each tested fungicide on *F. solani*.

4. Talk about the necessity of the toxicity test of pesticides *in vitro*.

第三部分　普通植物病理学实习
PART THREE　Practice of General Plant Pathology

实习一　植物病害的田间调查

【概述】

植物病害调查是指实地了解植物病害种类、分布、发生程度及其发生规律，并采集样本或收集相关资料。此项工作是开展植物病害预测预报、损失估计和有效防治不可缺少的基础性工作。

【目的和要求】

通过对植物病害的调查，了解植物病害种类、分布和发生情况，掌握植物病害的发生规律；学习植物病害调查和统计的基本方法，学会计算发病率和病情指数，为学习病害的预测预报和防治奠定基础。

【材料和用具】

1. 植物病害

田间或林间的植物病害；杨树溃疡病和苹果褐斑病。

2. 调查用具

手持放大镜、笔记本、铅笔、标签、刻度尺等。

【内容和方法】

植物病害调查一般分为普查和重点调查，可根据不同的调查目的选择合适的调查方法。

1. 调查类型

①普查：当一个地区有关植物病害发生情况的资料很少时，可先进行普查，主要是了解植物病害的分布特点、分布类型和发病程度，调查的面积要广，并且具有代表性。

②重点调查：经普查发现的重要病害，可对其分布、发病率、损失率、生态环境和防治效果等进行深入调查，可采取定时、定点和设标准地进行系统观察记录，调查数据比较准确。所以重点调查其目的性、实用性较强，可为植物病害的预测、预报和防治提供科学依据，在植物病理学研究中较为常用。

2. 调查内容和记录方法

调查的内容根据调查目的而定，一般包括植物病害的种类、分布、栽培管理与病害发

表 3-1 植物病害调查报告表

<table>
<tr><td colspan="5">基本信息</td></tr>
<tr><td colspan="3">寄主名称：</td><td colspan="2">病害名称：</td></tr>
<tr><td colspan="3">调查日期：</td><td colspan="2">调查地点：</td></tr>
<tr><td colspan="3">调查人：</td><td colspan="2">调查单位：</td></tr>
<tr><td colspan="3">种苗来源：</td><td colspan="2">土壤性质：</td></tr>
<tr><td colspan="3">管理措施：</td><td colspan="2">当地气候条件：</td></tr>
<tr><td colspan="3">施肥用药情况：</td><td colspan="2">当地群众对病害的认知：</td></tr>
<tr><td rowspan="7">病害分级标准</td><td>分级</td><td>病情严重程度描述</td><td>代表值</td><td>调查植株(叶片)数</td></tr>
<tr><td>Ⅰ</td><td></td><td>0</td><td></td></tr>
<tr><td>Ⅱ</td><td></td><td>1</td><td></td></tr>
<tr><td>Ⅲ</td><td></td><td>2</td><td></td></tr>
<tr><td>Ⅳ</td><td></td><td>3</td><td></td></tr>
<tr><td>Ⅴ</td><td></td><td>4</td><td></td></tr>
<tr><td>Ⅵ</td><td></td><td>5</td><td></td></tr>
<tr><td colspan="5">调查植株(叶片)总数：</td></tr>
<tr><td rowspan="2">调查结果</td><td colspan="4">发病率：</td></tr>
<tr><td colspan="4">病情指数：</td></tr>
<tr><td>备注</td><td colspan="4"></td></tr>
</table>

生的关系等，应将调查的原始数据详细地记载于调查报告，见表 3-1。

3. 植物病害田间调查的时间和次数

在实际工作中，往往根据调查目的及植物病害的发生特点确定调查的时间与次数。如苹果锈病每年都有发生，但是每年发生的轻重程度不同。为了解苹果锈病的发生与流行规律，可在苹果展叶期至落叶期(一般为每年的 3~10 月)每隔 15 d 进行一次调查，可连续调查 3~4 年，详细记录各项调查数据并进行分析，明确该病害的流行规律，为确定病害防治的最佳时间提供依据。

4. 植物病害调查的取样方法

病害调查的取样方法直接影响调查结果的准确性。取样方法应根据病害种类和环境条件而定：气流传播而分布均匀的病害，如小麦锈病、花椒落叶病等，取样点可以少些；土壤传播的病害，如棉花枯萎病、花椒根腐病等，样点要多些。凡地形、土壤耕作不一致的地区，应分别取样。为了使调查样点具有代表性，要根据病害种类、病害分布选用合适的方法。常用的取样方法如图 3-1 所示。

①五点随机取样法：设 5 个取样点，各点做到随机取样，此法适宜于分布均匀病害的调查，最为常用。

②棋盘式取样法：此方法准确性较高，但费时，适用于对土传病害、聚集分布型或负二项分布型病害的调查。

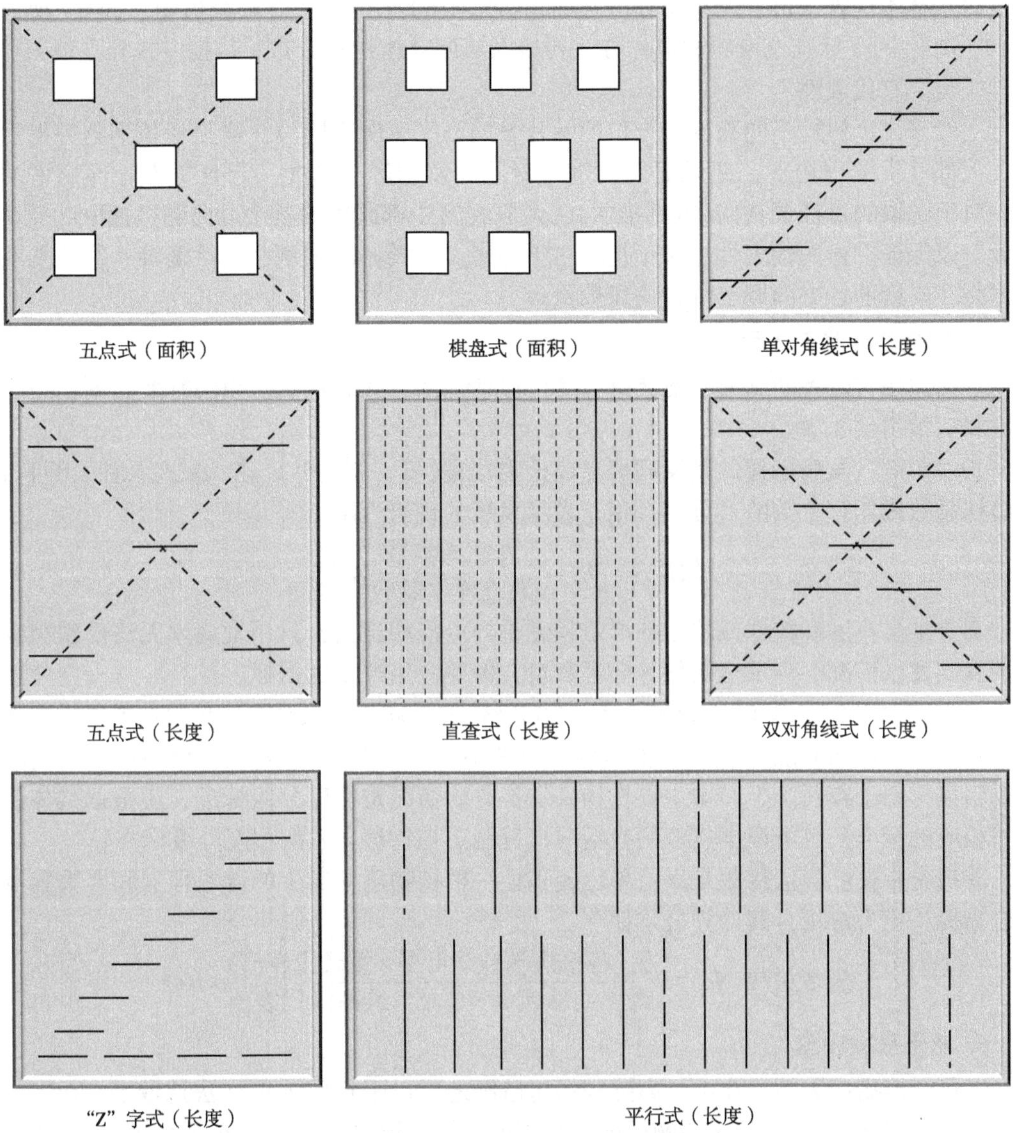

图 3-1　植物病害调查取样方法示意图

③对角线式取样法：进一步分单对角线和双对角线两种取样方法，适用于面积较大的近方形地块的病害调查，适合于气流和种子传播的病害调查。

④“Z”字式取样法：适于狭长地形或标准地边缘带病害分布较多的地块的病害调查。

⑤平行式取样法：间隔一定行数进行取样调查，此法适宜于苗圃、果园林木病害调查。

取样单位应随植物种类和病害特点而相应变化，一般以面积(密植作物)或长度(密植条播作物)为单位，也可以植株或器官为调查单位(果树等稀植作物)。取样数量取决于病害的发生特点，发病均匀的，取样量可少些，不均匀的要多些。在每一调查点上取 100~200 株或 20~30 个器官(如叶片)。取样量越大，越具有代表性，精确度越高，但也较费时费工。

对于林木病害来讲，调查实际取样面积(或株树)不得小于被调查植物总面积(或总株数)的5%。调查标准地设置和取样方法要根据被调查植物的分布、栽植方式和立地条件而定，一定要有代表性。

一般来说，树冠不同高度、方位的叶片发病程度有所不同，往往呈现顶部新叶发病较轻、下部叶片发病较重；迎风面发病重，背风面发病轻。因此，调查树木叶部病害病情时，对于选取的每株调查树(或标准木)，应在树冠中部高度的各个方向随机抽取叶片调查统计，以保证取样的代表性。对于树木主干病害，如杨树溃疡病、腐烂病等，往往要在标准木胸径高度的各方向调查统计病斑数量。

5. 病情的计算与表示法

对田间病害调查的结果，要经过计算，以适当的方式表示。常用的有发病率(%)、严重度和病情指数3项。

①发病率：发病率指的是发病植株或植物器官(根、茎、叶、花、果实、种子等)占调查植株总数或器官总数的百分率，用于表示发病的普遍程度。

$$发病率(\%)=\frac{发病样本数}{调查样本总数}\times 100$$

②严重度：发病植株或器官的严重程度差异可能相当大，只用发病率无法反映植物病害的真实发生情况。为了全面估计病害数量，需要应用严重度指标。

严重度常用0~9或0~5级来表示，“0”级代表无症状，最高数值代表最严重的发病程度，然后以一定的间隔，分为若干级别，级别不需要太多，一般3~5级即可。级别间差异要明显，便于判断。生产中主要作物病害都有病情严重程度分级标准，应根据具体病害查找相应的标准，也可根据实地调查情况自行制定植物病害发病程度分级标准。

③病情指数：病情指数是反映单位面积上一种植物病害发生的普遍程度和严重程度的综合指标，更为常用，其计算公式如下：

$$病情指数(\%)=\frac{\sum(各级病株或病叶数\times 各级代表值)}{调查病株或病叶总数\times 最高级代表值}\times 100$$

6. 枝干病害调查

以杨树溃疡病为例，在中、高感病杨树苗圃地，以平行线式取样法选取5~10年生标准木50株左右，测量各杨树的胸径值，并用记号笔标记出距离胸径位置为20 cm的树干高度范围，统计该范围内树皮上的溃疡病斑数。

先按照以下公式计算出各标准木的单位胸径病斑数，再按照该病害发生程度分级标准(表3-2)计算病情指数。

$$单位胸径病斑数\ N(个/\mathrm{cm})=\frac{20\ \mathrm{cm}\ 高度范围内病斑总数(个)}{标准木胸径值(\mathrm{cm})}$$

表3-2　杨树溃疡病分级标准

严重度分级	发病程度	代表数值	严重度分级	发病程度	代表数值
Ⅰ	无病斑，$N=0$	0	Ⅳ	$5<N\leqslant 10$	3
Ⅱ	$0<N\leqslant 1$	1	Ⅴ	$10<N\leqslant 15$	4
Ⅲ	$1<N\leqslant 5$	2	Ⅵ	$N>15.0$	5

7. 叶片病害调查

以苹果叶片褐斑病为例，在苗圃地采取平行式取样方法随机(在植株的东南西北和上中下多方位)摘取苹果叶片，每个取样点不少于500个叶片，根据实际调查情况计算发病率，制定苹果褐斑病分级标准(表3-3)，计算病情指数。苹果褐斑病的发病程度用叶片病斑面积占叶片面积的比例(S)来表示。

表3-3 苹果褐斑病分级标准

严重度分级	发病程度	代表数值	严重度分级	发病程度	代表数值
Ⅰ	无病斑，$S=0$	0	Ⅳ	$1/3<S\leqslant1/2$	3
Ⅱ	$0<S\leqslant1/5$	1	Ⅴ	$1/2<S\leqslant2/3$	4
Ⅲ	$1/5<S\leqslant1/3$	2	Ⅵ	$2/3<S\leqslant1$	5

【作业和思考题】

1. 选取一种重要的植物病害，以小组为单位对其发病情况进行调查，详细记录调查情况，自行制定该病害的严重度分级标准，并计算该病害的发病率和病情指数。

2. 谈谈进行植物病害调查的意义。

PRACTICE 1 Field Investigation of Plant Diseases

【Introduction】

Plant disease investigation is to understand the extent of plant disease distribution and its occurrence rules, and collect samples or relevant data. This work is an indispensable basic work for plant disease prediction, loss estimation and effective disease control.

【Purpose and Requirements】

Understand the types, distribution and occurrence of plant diseases and grasp the occurrence rules of plant diseases through the investigation of plant diseases. Learn the basic methods of plant disease investigation and statistics and learn to calculate the incidence of disease and disease index, which lay the foundation for the study of disease prediction and prevention.

【Materials and Appliances】

1. Materials

The plant diseases in the field or woodland, poplar canker and apple brown spot.

2. Tools

Hand-held magnifying glasses, notebooks, pencils, labels, scales, etc.

【Contents and Methods】

Plant disease investigation is generally divided into census and key investigation, and appropriate investigation methods can be selected according to different investigation purposes.

1. Types of Investigation

①General investigation: When there is little information about the occurrence of plant diseases in an area, a census can be carried out first, mainly to understand the distribution characteristics of plant diseases, distribution types and incidence degree. The survey area should be wide and representative.

②Key investigation: For important diseases discovered by census, in-depth investigation can be carried out on their distribution, incidence, loss rate, ecological environment, and prevention effect, etc. Systematic observation and record can be carried out at fixed time, fixed site, and standard places, and the survey data are more accurate. Therefore, the focus investigation has strong purpose and practicability. It can provide scientific basis for the prediction, forecast, and prevention of plant diseases, and is commonly used in the study of plant pathology.

2. Investigation Contents and Recording Methods

The contents of the survey are determined according to the purpose of the survey, which generally includes the relationship between the distribution and types of plant diseases, cultivation management, and disease occurrence, etc. The original data of the survey should be recorded in the survey report in detail, as shown in Table 3-1.

3. Time and Frequency of Field Investigation of Plant Diseases

In practice, the time and times of investigation are often determined by the purpose of investigation and the occurrence of plant diseases. For example, apple rust occurs every year, but the severity is not the same. To understand the occurrence and epidemic of apple rust, a survey can be conducted every 15 d during the exhibition of apple leaf to deciduous(usually 3-10 month of each year), and the survey should be continuous carried out for 3-4 years. A detailed record of various investigation data and analysis, and the epidemic regularity of the disease, can provide guidance to determine the best time for disease control.

4. Sampling Methods for Plant Disease Investigation

The sampling method of disease investigation directly affects the accuracy of investigation. Sampling methods should be determined according to disease types and environmental conditions. For diseases with uniform distribution due to airflow, such as wheat rust and deciduous disease of *Zanthoxylum bungeanum*, sampling points sites can be fewer; While for soil-borne, such as cotton wilt and root rot of *Z. bungeanum*, more sample points are necessary. Areas with inconsistent topography and soil cultivation should be sampled separately. In order to make the survey sample points representative, appropriate methods should be selected according to the species and distribution of the disease. Commonly used sampling methods are shown in Figure 3-1.

Table 3-1 Survey Report Table of Plant Diseases

<table>
<tr><td colspan="5">Basic information</td></tr>
<tr><td colspan="3">Host name:
Survey date:
Investigators:
Source of seeds or seedlings:
Managements:
Application of fertilizers and pestcides:</td><td colspan="2">Disease name:
Survey site:
Survey organization:
Soil properties:
Local climatic conditions:
Local awareness of disease:</td></tr>
<tr><td rowspan="7">Disease classification standard</td><td>Grade</td><td>Disease severity</td><td>Representative value</td><td>Total number of surveyed plants(leaves)</td></tr>
<tr><td>Ⅰ</td><td></td><td>0</td><td></td></tr>
<tr><td>Ⅱ</td><td></td><td>1</td><td></td></tr>
<tr><td>Ⅲ</td><td></td><td>2</td><td></td></tr>
<tr><td>Ⅳ</td><td></td><td>3</td><td></td></tr>
<tr><td>Ⅴ</td><td></td><td>4</td><td></td></tr>
<tr><td>Ⅵ</td><td></td><td>5</td><td></td></tr>
<tr><td colspan="5">Total number of surveyed plants(leaves):</td></tr>
<tr><td rowspan="2" colspan="2">Survey results</td><td colspan="3">Disease incidence:</td></tr>
<tr><td colspan="3">Disease index:</td></tr>
<tr><td colspan="2">Note</td><td colspan="3"></td></tr>
</table>

①Five-point random sampling method: Five sampling points are set up, and random sampling is done at each point. This method is suitable for the investigation of evenly distributed diseases and is the most commonly used method.

②Checkerboard sampling method: This method is more accurate, but time-consuming. It is suitable for the investigation of soil-borne diseases with clustered distribution or negative binomial distribution.

③Diagonal sampling method: It is further divided into two sampling methods, single diagonal and double diagonal. It is suitable for the disease investigation of nearly square plots with large area, and airborne and seed-borne disease.

④Z-line sampling method: It is suitable for the disease investigation of long and narrow terrain or the plot with more disease distribution in the edge zone of standard land.

⑤Parallel sampling method: Sampling investigation with a certain number of rows at intervals, which is suitable for the investigation of tree diseases in the nursery and orchard.

The sampling unit should be changed according to the plant species and disease characteristics, generally taking the area(dense planting crops) or length(densely drill planting crops) as the unit, or the plant or organ as the investigation unit (sparse crops such as fruit trees). The number of samples depends on the characteristics of the disease. If the disease is uniform, the number of samples can be less, or else the number of samples should be more. 100-

200 plants or 20-30 organs (such as leaves) will be taken from each investigation site. The larger the sample size, the more representative and accurate it is, but it is also more time-consuming and labor-intensive.

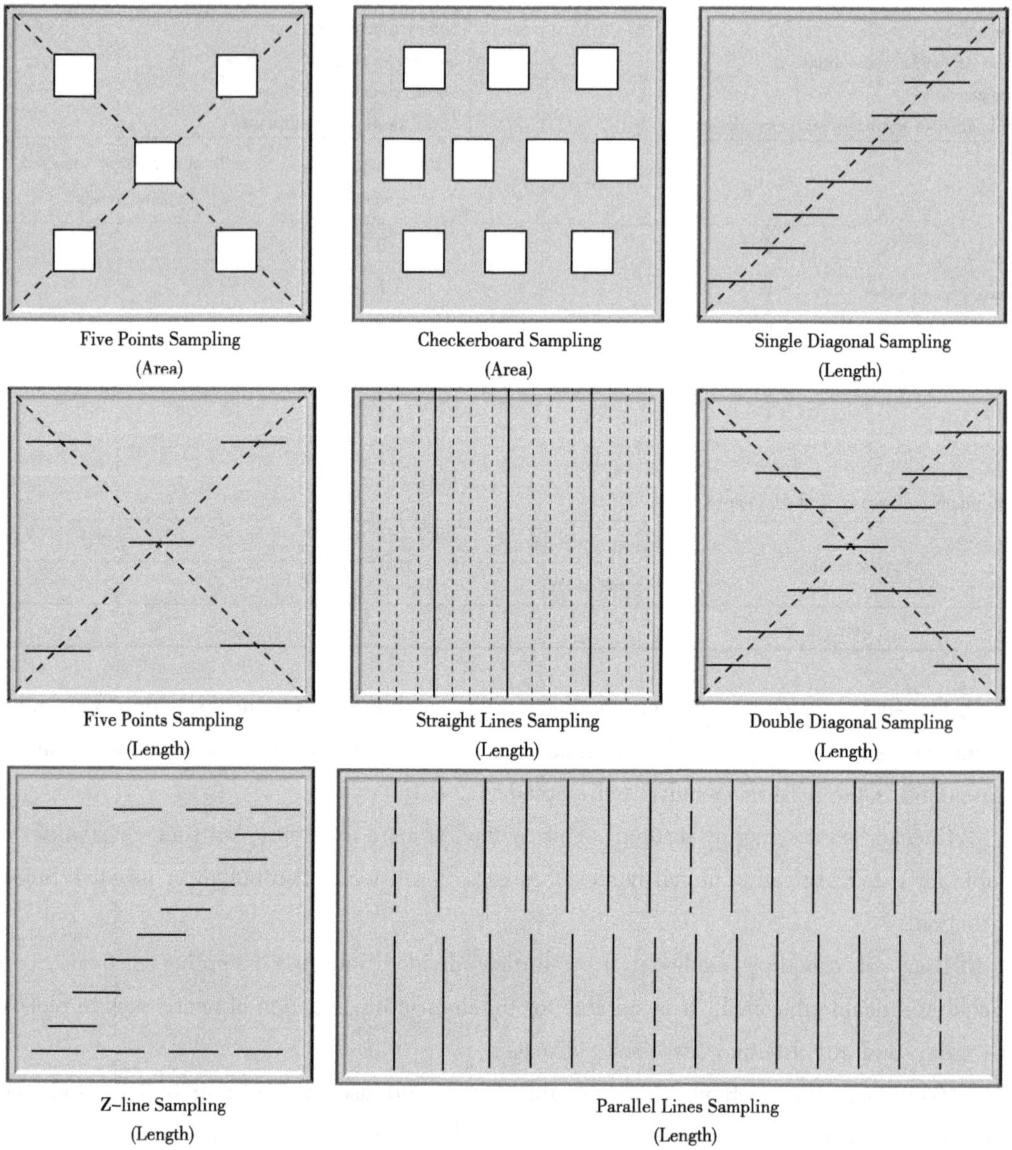

Figure 3-1 Schematic Diagram of Sampling Methods for Plant Disease Investigation

For forest tree diseases, the actual sampling area (or trees) should not be less than 5% of the total area (or total number of plants) of the investigated plants. The standard site setting and sampling method of the survey should be determined according to the distribution, cultiviation, and soil condition of the investigated plants, and must be representative.

Generally, the degrees of disease on leaves in the top and bottom of the tree crown and in the east, west, north, and south directions are different. Normally, the disease of new leaves at the top is usually lighter and that at the bottom is more severe. The disease of leaves at upwind side is heavy, and that at back side is light. Therefore, in the investigation of tree leaf diseases, for each selected investigation tree(or standard tree), the leaves should be randomly selected from all directions of the middle height of the tree crown for investigation and statistics to ensure the representativeness of the sampling. For the diseases of the stem of tree, such as poplar stem canker and rot, the number of disease spots should be counted in all directions of the height of the standard diameter at breast height(DBH).

5. Calculation and Representation of Plant Disease Condition

The results of field disease investigation should be calculated, and the three commonly used parameters, including incidence (%), severity, and disease index, should be expressed in appropriate ways.

①Incidence: It refers to the percentage of infected plants or plant organs (roots, stems, leaves, flowers, fruits, seeds, etc.) in the total number of investigated plants or organs, which is used to indicate the general condition of plant disease.

$$\text{Incidence}(\%)=\frac{\text{Number of infected samples}}{\text{Total number of investigated samples}}\times 100$$

②Severity: The severity of infected plants or organs may vary considerably, and incidence alone cannot reflect the true occurrence of plant disease. In order to comprehensively estimate the number of diseases, severity indicators should be applied.

Severity is usually expressed as 0-9 or 0-5 grade, Grade 0 represents asymptomatic, and the highest value represents the most serious degree of disease. Then, at a certain interval, it is divided into several levels, which do not need too many levels. Generally, 3-5 levels are sufficient. The differences among different levels must be obvious and easy to judge. Major crop diseases in production have disease severity grading standards. The corresponding standards should be found according to the specific diseases, or the grading standards of disease severity of plant diseases can be made by themselves according to the field investigation.

③Disease index: It is a comprehensive index reflecting the prevalence and severity of the occurrence of a plant disease per unit area. It is more commonly used. The calculation formula is as follows:

$$\text{Disease index}(\%)=\frac{\sum(\text{Number of infected samples}\times\text{Disease grade})}{\text{Number of total infected samples}\times\text{Highest grade}}\times 100$$

6. Investigation of Stem Disease

Taking poplar canker as an example, select about 50 standard woods at 5-10 years in the medium and high susceptible poplar nursery according to parallel type sampling method, and measure the diameter at breast height(DBH) of each poplar wood. A tree trunk height range of 20 cm from DBH was marked with a marker 20 cm height, and count the number of canker spots locating this range.

First, the number of spots per DBH of each standard wood is calculated according to the following formula, and then the disease index was calculated according to the grading standard (Table 3-2).

$$\text{Number of spots per DBH}(N) = \frac{\text{Number of spotsat 20 cm height}}{\text{DBH of tree(cm)}}$$

Table 3-2　Grading Standard of Poplar Canker

Severity grade	Degree of disease	Representative value	Severity grade	Degree of disease	Representative value
Ⅰ	No spots, $N=0$	0	Ⅳ	$5<N\leq 10$	3
Ⅱ	$0<N\leq 1$	1	Ⅴ	$10<N\leq 15$	4
Ⅲ	$1<N\leq 5$	2	Ⅵ	$N>15.0$	5

7. Leaf Disease Investigation

Taking apple brown spot as an example, collect the apple leaves randomly at different directions (east, south, west, and north) and sites (top, middle, and bottom) in the nursery according to parallel random sampling method. Not less than 500 leaves are collected at each sampling point. Calculate the incidence based on the actual investigation and make a grading standard of apple brown spot (Table 3-3) to calculate the disease index. The degree of apple brown spot is expressed using the ratio of leaf lesion area to leaf area (S).

Table 3-3　Grading Standard of Apple Brown Spot

Severity grade	Degree of disease	Representative value	Severity grade	Degree of disease	Representative value
Ⅰ	Symptomless, $S=0$	0	Ⅳ	$1/3<S\leq 1/2$	3
Ⅱ	$0<S\leq 1/5$	1	Ⅴ	$1/2<S\leq 2/3$	4
Ⅲ	$1/5<S\leq 1/3$	2	Ⅵ	$2/3<S\leq 1$	5

【Assignments and Questions】

1. Select an important plant disease and investigate its incidence in a group, record the investigation in detail, make the severity classification standard of the disease, and calculate the disease incidence and disease index.

2. Discuss the significance of plant disease investigation.

实习二　植物病害标本的采集、制作与保存

【概述】

植物病害标本与其他生物标本一样，是生物性状和分布的记载。一份好的植物病害标本就是最完整的症状描述和性状特征的记载，因此，采集、制作和保存植物病害标本，在

普通植物病理学的教学和科研中都是非常重要的。

【目的和要求】

1. 了解植物病害标本的采集要求，掌握植物病害标本的制作与保存方法。

2. 每小组采集并制作 80~100 号植物病害标本。

【材料和用具】

1. 实习材料

田间或林间的各种植物病害。

2. 实习用具

标本夹、标本纸、采集袋、采集盒、采集标签、塑封袋、采集记录本、手持放大镜、剪枝剪、手锯、高枝剪、电工刀、手持 GPS、相机、铅笔、记号笔、标本缸、标本瓶、95%乙醇、甲醛、硫酸铜、亚硫酸、醋酸铜、氯化锌、甘油、蒸馏水等。

【内容和方法】

1. 植物病害标本的采集

采集植物病害标本时，需要综合考虑采集地的地理生态条件、病原物的生物学特性和植物种类等因素与植物病害发生的关系。

①标本的采集要求：植物病害标本要能表现出植物病害的特征，须注意采集病害症状的各种表现。对不认识的寄主植物，一定要采集其花、果实以备鉴定，对常见的局部性病害寄主，就不一定采植物整株。但若植物各个器官都发病，则应分别采集各个发病器官。病征是鉴定植物病害的主要依据，凡有病征的植物病害，一定要注意采集带有典型病征或病原物不同发育阶段繁殖体的标本。

②异常植物的观察：采集时，首先应对一种植物的异常表现进行观察，主要观察它是病害、虫害还是机械伤害？是侵染性病害还是非侵染性病害？这种病害特点是什么？哪些病株可以代表这种病害的特点？

③标本的采集：选定具有典型症状的发病植物后，采集症状明显的发病组织部位。每种标本采集的份数不能太少，一般要求每种标本采 3~5 份，因为在制作和鉴定的过程中常有损坏，务必保证还有多余的标本可用于保存和交流。

④标本的编号记录：当采集到一定数量的植物病害标本后，即可对其进行编号记录。按照采集顺序给采集的标本指定“标本号”。采集标签和采集记录本是记载标本症状特点的重要备查依据，应及时详细填写。

2. 植物病害标本的制作与保存

从野外采集的植物病害标本，除一部分用于鉴定外，对于症状典型的标本最好先进行摄影后，再压制或浸渍保存。压制或浸渍保存的标本应尽可能保持其原有性状，微小的标本可以制成玻片，如双层玻片、凹穴玻片或用其他小玻瓶和小袋收藏。

(1) 干燥压制法

此法简单、经济，标本可以长期保存，所以应用最广。将采集的标本展平后夹在吸水的标本纸中，同时放入采集标签，用标本夹压紧后日晒或烘烤，使其干燥，干燥越快越能保持标本的原有色泽。标本质量越高的干燥标本才能成为有价值的材料永久保存。为此务必做到：

①在压制前须进行标本的修剪和整形，然后放在标本纸中加压吸水干燥。

②勤换纸、翻压标本。新采的标本，每日至少更换标本纸一次，遇秋季高湿及多汁标本，每日应换两次。

③初次换纸翻压时应对标本进行整形，3~5 d 后，标本已大部分干燥，可以隔日换纸、翻压直至干燥。

(2)普通防腐浸渍法

对于多汁的植物病害标本，如果实、块根或担子菌的子实体等，必须采用浸渍法保存。浸渍液种类较多，有纯防腐性的，也有专为保持标本原来色泽的。普通防腐浸渍法仅能防腐而没有保持标本原色的作用，如花椒根结线虫病、大豆胞囊线虫病等标本可采用此方法保存。将洗净后的标本直接浸于 5%甲醛浸渍液或亚硫酸浸渍液(1 000 mL 水中加入 5%~6%的亚硫酸溶液 15 mL)。

(3)保存绿色标本的浸渍法

①醋酸铜浸渍法：以结晶醋酸铜逐渐添加到 50%的醋酸中，至饱和为止，使用时兑水稀释 1~4 倍(稀释倍数视标本颜色而异，色深者稀释倍数可小些)。该方法又可分为热处理法和冷处理法。

热处理法：将稀释后的溶液加热至沸腾，放入标本，标本的绿色最初会被漂去，经 3~4 min 后标本会恢复绿色，此时，将标本取出，用清水漂净，保存于 5%甲醛中。

冷处理法：将标本放入稀释后的浸渍液中，浸泡 3 h 左右至标本褪色，再经 72 h 后标本又恢复绿色，此时将标本取出，用清水漂净，保存于 5%甲醛中。此法保持色泽时间较长，其保色原理大致是铜离子与叶绿素中的镁离子发生置换作用。所以，溶液经多次处理标本后，铜离子会逐渐减少。如要继续使用，应补充适量的醋酸铜。此法保存的标本往往略带蓝色，与植物标本原色稍有差异。

②硫酸铜—亚硫酸浸渍法：用清水冲净的标本，直接浸泡在 5%硫酸铜溶液中 1~7 d，待标本略带褐色时取出用清水冲净标本表面的硫酸铜溶液，然后保存于亚硫酸浸渍液中。

(4)保存黄色和橘红色标本的浸渍法

含有叶黄素和胡萝卜素的果实，如杏、红辣椒等病害标本可用亚硫酸浸渍液保存。不同的标本使用浸渍液浓度应该经过试验确定。亚硫酸浓度过高，有漂白作用，不能保色；浓度过低，防腐力不足。如确需使用低浓度的亚硫酸浸渍液时，可添加适量的乙醇增加防腐能力。此外，为了防止果实崩裂，可添加少许的甘油。

(5)保存红色标本的浸渍法

红色标本的保存较困难，因为红色是由花青素形成的，花青素能溶于水和乙醇。常用以下几种浸渍液保存红色标本：

①Hesler 浸渍液：其成分为氯化锌 50 g，甲醛 25 mL，甘油 25 mL，加水定容至 1 000 mL。将氯化锌溶于热水中，加入甲醛和甘油，如有沉淀，用其澄清液。

②波尔浸渍液：将二氧化硫通过甲醛得到饱和溶液，使用时稀释 20~40 倍。

③瓦查保存红色浸渍液：其配方是硝酸亚钴 15 g，氯化锡 10 g，甲醛 25 mL，加水定容至 2 000 mL。常用于保存草莓、辣椒、马铃薯及其他组织中的红色。标本洗净后，在此浸渍液中浸泡 14 d(完全浸没)，然后保存在以下浸渍液中，其配方为：甲醛10 mL，95% 乙醇 10 mL，饱和亚硫酸溶液 30~50 mL，加水定容至 1 000 mL。

(6)浸渍标本的封口

浸渍标本的封口很重要，因为配置浸渍液所用的药品大都具有挥发性，或者是容易氧化。密封可以保持浸渍液的效果，封口方法如下：

①临时封口法：将蜂蜡和松香各 1 份，分别融化后混合，加入少量凡士林调成胶状，涂在瓶盖边缘，将盖压紧封口。明胶和石蜡的混合物也能用于封口，将明胶(4 份)在水中浸泡几小时，滤水后加热融化，加入石蜡 1 份，融化后即成为胶状物，趁热使用。

②永久封口法：将酪胶和消石灰各 1 份混合，加水调成糊状，即可用于封口。干燥后，因酪酸钙的硬化而密封。此外，将明胶 28 g 在水中浸泡几个小时，滤水后加热融化，加入重铬酸钾 0. 324 g 和适量的熟石膏调成糊状，也可用于永久封口。

3. 标本的整理和保存

制作好的标本，经整理和登记后，将其按一定的顺序进行排列和保存。常用的保存方法有以下几种：

①标本盒保存：教学和示范的病害标本，除浸渍标本外，用玻面标本盒保存比较方便。玻面标本盒大小不一，适宜的大小是 200 mm×280 mm×(15~30)mm。标本盒中先铺一层棉花，棉花上放标本和标签，注明寄主植物和病原物名称，盖上玻盖。棉花中可加入少许樟脑粉或其他驱虫剂。

②标本瓶保存：浸渍的标本放在标本瓶内保存，为避免标本的漂浮和移动，可将标本固定在玻璃条上。盖好瓶盖并封口，贴上标签。标本瓶应放置在暗处，以减少药液的氧化，或瓶口因温度变化太大而碎裂。

③牛皮纸信封袋保存：将压制好的标本放在牛皮纸信封袋内，信封袋的规格依标本大小而定，然后将鉴定标签贴在信封袋外面，然后放入标本柜。大量保存的干制标本，常采用此方法保存。

④塑封保存：将标本平整夹于塑料内，通过塑封机加热排净空气并密封。

【作业和思考题】

1. 每小组采集有效植物病害标本 80~100 号，并提交其压制或浸渍标本。
2. 什么是有效的植物病害标本?
3. 如何选择植物病害标本的制作方法?
4. 谈谈植物病害标本采集与保存的重要性。

PRACTICE 2 Collection, Preparation and Preservation of Plant Disease Specimens

【Introduction】

As same as the other biological specimen, plant disease specimen is the record of biological character and distribution. A good plant disease specimen is the record of complete description of symptoms and characteristics. Therefore, the collection, preparation and preservation of plant disease specimens are very important in teaching and scientific research of general plant pathology.

【Purpose and Requirements】

1. Understand the collection requirements of plant disease specimens, and master the methods of preparation and preservation of plant disease specimens.

2. Collect and prepare 80-100 kinds of specimens for each group.

【Materials and Appliances】

1. Materials

Various plant diseases in the field or forest.

2. Tools

Specimen holders, specimen papers, collection bags, collection boxes, collection labels, plastic bags, collection record books, hand-held magnifying glasses, pruning shears, hand-held saws, high-branch shears, electrician knives, hand-held GPS, cameras, pencils, marking pens, specimen jars, specimen bottles, 95% alcohol, formaldehyde, copper sulfate, sulfurous acid, copper acetate, zinc chloride, glycerin, distilled water, etc.

【Contents and Methods】

1. Collection of Plant Disease Specimen

It is necessary to comprehensively consider the relationship between the occurrence of plant disease and the factors, i. e. the geographical and ecological conditions of the collection place, the biological characteristics of the pathogens and the types of plants when collecting plant disease specimens.

①Requirements for specimen collection: Specimens of plant disease need to be able to show the characteristics of plant disease. Attention must be paid to the collection of various manifestations of disease symptoms . For unknown host plants, collect their flowers and fruits for identification. For common localized disease hosts, it is not necessary to collect the whole plant. But if all plant organs are infected, each infected organ should be collected. Signs are the main basis for the identification of plant diseases. For any plant diseases with signs, attention must be paid to the collection of specimens with typical signs or propagules of pathogens in different developmental stages.

②Observation of abnormal plants: The abnormal exhibition of a plant should be observed when collecting, and mainly observe whether it is caused by diseases, insects or mechanical damages? Is it an infectious disease or a non-infectious disease? What is characteristic of this disease? Which diseased plants can represent the characteristics of this disease?

③Collection of specimen: The diseased tissues with obvious symptom are collected after the typical diseased plants is selected. The number of each specimen should not be too few.

Generally, it is required to collect 3-5 copies of each specimen. Because damage is often happened in the process of production and identification, it is necessary to ensure that there are enough specimens available for preservation and communication.

④Numbering record of specimen: The plant disease specimens can be record by numbering after collecting a certain amount of specimens. The collected specimens are assigned the "specimen number" according to the collection sequence. Collection label and record book are the important basis for recording the symptoms and characteristics of specimens, and should be filled in time and in detail.

2. Preparation and Preservation of Plant Disease Specimen

Except for a part for identification, the specimens collected from the field with typical symptoms should be photographed first, and then preserved using the methods of pressing or immersing. It's better to keep the original character of the specimens prepared by pressing or immersing. Tiny specimens can be made into glass slides, such as double glass slides, cave slides, or collection with other small glass bottles and pouches.

(1) Dry Pressing of Specimen

This method is simple and economic, and the specimen can be preserved for a long time, so it is the most widely used method. The specimens are flattened and placed onto specimen papers, while the collection lables are placed together. Press the specimen holder and then bask or bake for drying. The faster the drying, the better it can maintain the original color of the specimen. Only the dry specimen with high quality can become a valuable material for permanent preservation. Therefore, it is important to do:

①Before pressing, the specimen must be trimmed and reshaped, and then placed in specimen paper to press for water absorption and drying.

②Frequently change specimen paper, turn over specimens. For newly collected specimens, dry specimen paper should be changed at least once a day. In case of high humidity in autumn and juicy specimens, it should be changed twice a day.

③The specimen should be reshaped when the paper is changed and turned over for the first time. After three or five days, most of the specimen is dry, so the paper can be changed and turned over every two days until it is dry.

(2) Ordinary Antiseptic Immersing Method

It's necessary to apply immersing method to preserve the juicy plant disease specimens, such as fruits, root tubers, or fruiting bodies of Basidiomycota, etc. There are a variety of impregnation solution, such as anticorrosive only, or that designed for color maintaining. The ordinary antiseptic immersing method can only prevent corrosion without keeping the original color of specimens. For example, the specimens of root knot nematode of *Zanthoxylum bungeanum* and soybean cystic nematode can be preserved by this method. The washed specimens can be directly immersed in 5% formalin impregnation solution or sulfurous acid impregnation solution (add 15 mL of 5%-6% sulfurous acid solution into 1 000 mL water).

(3) Immersing Method for Green Specimens Preservation

①Copper acetate immersing method: Add crystalline copper acetate gradually to 50% acetic acid until saturation, and dilute it with water 1-4 times for use (the dilution ratio varies according to the color of the specimen, and the dilution ratio can be smaller for the darker color). This method can be divided into heat treatment method and cold treatment method.

Heat treatment method: Heat the diluted solution to boiling and put specimen into the solution. The green color of the specimen would be washed away initially, and be recovered after 3-4 min. At this time, take out the specimen, wash it with clear water and store it in 5% formalin.

Cold treatment method: Dip specimen in the diluted solution, soak it for about 3 h to fade, after 72 h the specimen turns to green again. Remove the specimen, wash it with clean water, preserve it in 5% formalin. This method maintains the color for a long time. The general principle of color preservation is copper ions replace magnesium ions in the chlorophyll. Therefore, after the solution has been used for many times, the copper ions gradually decrease. If you want to continue to use the solution, an appropriate amount of copper acetate should be added. The specimens preserved by this method are often bluish, which is slightly different from the original color of plant specimens.

②Copper sulfate and sulfurous acid immersing method: Wash the specimen with clean water, soak it directly in 5% copper sulfate solution for 1-7 d. When the specimen is slightly brown, take it out and wash off the remaining copper sulfate solution on the surface and then store it in the sulfurous acid impregnation solution.

(4) Immersing Method for Yellow and Orange Specimens

Fruit containing lutein and carotene, such as apricot red pepper and other disease specimens can be preserved with sulfurous acid impregnation solution. The concentration of the impregnation solution should be determined. Too high concentration of sulfurous acid has a bleaching effect, and cannot retain color; if the concentration is too low, the anti-corrosion ability is weak. If it is really necessary to use a low concentration of sulfurous acid impregnation solution, appropriate amount of alcohol can be added to increase the antiseptic ability. In addition, in order to prevent the fruit from cracking, a little glycerin can be added.

(5) Immersing Method for Red Specimens

Red specimens are more difficult to preserve because red color is formed by anthocyanins, which can dissolve in water and alcohol. The following impregnation solutions are commonly used to preserve red specimens:

①Hesler impregnation solution: The formula is 50 g zinc chloride, 25 mL formalin, and 25 mL glycerol in 1 000 mL. Dissolve zinc chloride in hot water, add formalin and glycerol, and use the clarifying solution if there is precipitation.

②Bohr impregnation solution: Pass sulfur dioxide through formalin to obtain a saturated solution, dilute 20-40 times for use.

③Vacha impregnation solution: The formula is 15 g cobaltous nitrate, 10 g tin chloride, 25 mL formalin, add water to the volume of 2 000 mL. It is often used for preservation of red specimens such as strawberry, pepper, potato, and other tissues. After washing, soak the specimens in this impregnation solution for 14 d(completely immersed), and then store them in the following impregnation solution, the formula of which is: 10 mL formalin, 10 mL of 95% alcohol, 30-50 mL saturated sulfite solution, add water to 1 000 mL.

(6) Sealing of Immersed Specimens

Sealing of impregnated specimens is important because most of the drugs used to prepare the impregnated solution are volatile or easily to oxidize. Sealing can maintain the effect of the impregnated solution. Sealing methods are as follows:

①Temporary sealing method: Melt wax and rosin(1 : 1), respectively, add a small amount of vaseline to make a gel. Smear the gel on the edge of the cap and press the cap tightly to seal. The mixture of gelatin and paraffin(4 : 1) can also be used to seal. Soak the gelatin in water for a few hours, filter, melt it by heating, and then add paraffin. Make a gel for use.

②Permanent sealing method: A mixture of casein and sluice lime(1 : 1) is added to make a paste, which can be used for sealing. After drying, the sealing is completed due to hardening of calcium casein. In addition, soak 28 g gelatin in water for several hours, melt, and add 0. 324 g potassium dichromate and appropriate amount of gypsum of paris to make a paste, which can also be used for permanent sealing.

3. Arrangement and Preservation of Specimens

The prepared specimens are arranged and preserved according to a certain order after sorting and registering. The common preservation methods are as follows:

① Preservation using specimen boxes: For the disease specimens for teaching and demonstration, except the impregnation specimens, it is more convenient to store them in a specimen box with the glass cover. Specimen boxes are in different sizes. The suitable size for the glass surface specimen box is 200 mm×280 mm×(15-30) mm. Lay a layer of cotton in the box first, put the specimens and labels on the cotton, mark the name of the host plant and the pathogen, and cover the box. A small amount of camphor powder or other insect repellent can be added in the cotton.

②Preservation using specimen bottles: The impregnated specimens should be preserved in the specimen bottles. In order to avoid the floating and moving, the specimen can be fixed on a glass strip. Cover the bottle and seal it. The specimen bottle should be placed in dark to avoid the oxidation of the liquid medicine, or the brokening of bottle due to too much temperature change.

③Preservation using kraft paper envelopes: Put the pressed specimen in a kraft paper envelope bag. The size of the envelope bag depends on the size of the specimen, attach the identification label outside of the envelope bag, and then put the bag into a specimen cabinet. This method is often used for preservation of dry specimens in bulk.

④Preservation by plastic sealing: Clamp the specimen evenly in plastic, heat it by a plastic sealing machine to drain the air and seal.

【Assignments and Questions】

1. Collect 80-100 valid plant disease specimens for each group and submit their pressed or impregnated specimens.

2. What is a valid plant disease specimen?

3. How to choose the preparation method for plant disease specimens?

4. Talk about the importance of collection and preservation of plant disease specimens.

实习三　植物病害的诊断与病原物的鉴定

【概述】

植物病害诊断是将植物病理学的基础知识和基本理论应用到农业生产实践中，对植物病害进行防治的前提。面对一种发病植物，首先要判断其是侵染性病害还是非侵染性病害。若是侵染性病害，则需要进行症状的全面观察、显微切片的制作与镜检、病原物的分离培养等一系列的操作方可鉴定出具体的病原物。

【目的和要求】

1. 了解不同类型植物病害的发生特点，掌握植物病害诊断和鉴定的基本方法。

2. 掌握植物病害诊断的具体步骤和制片方法，并能根据所学知识提出植物病害的合理防治方案。

【材料和用具】

1. 实习材料

采集的各种新鲜植物病害标本。

2. 实习器具

显微镜、显微成像系统、放大镜、载玻片、盖玻片、刀片、通草、镊子、乙醇、蒸馏水、吸水纸、擦镜纸、解剖针、铅笔、绘图纸、鉴定记录本、鉴定标签等。

【内容和方法】

1. 植物病害的诊断

植物病害通常可以根据其症状特点进行初步诊断。首先，根据植物病害的田间发病特征，初步判断其侵染性病害还是非侵染性病害。其次，仔细观察采集的植物病害标本的症状特点，并将其详细记录在鉴定记录本上。对病害症状特点明显的侵染性病害，根据各类病原物引起植物病害症状的特点，进一步确定其病原物的具体种(菌物、细菌、病毒或线虫等)。观察症状时，应注意以下事项：

①观察症状时，应先注意病害对全株的影响，然后检查病部。

②观察斑点病害时，要注意斑点的形状、数目、大小、色泽、排列和有无轮纹等；观察腐烂病害时，要注意腐烂组织的色泽、气味、结构(如软腐、干腐、湿腐)以及有无虫伤等。

③务必注意发病植物部位是否有病征，其对病原物的鉴定非常重要。一些菌物引起的植物病害通常会在病部出现粉、锈、霉、点、蕈菌等病征；一些细菌性植物病害会出现菌脓；病毒和植原体病害无病征。

④鉴定过程中可能会遇到很多非侵染性病害，它们的症状有时与病毒病、线虫病或有些真菌病害很相似，甚至在病部还能检查到真菌，当然，这些真菌一般是植物遭受非侵染性病害危害后，在坏死组织上生长的腐生菌。

⑤非侵染性病害在田间是普遍发生的，往往没有明显的发病中心，其与地形、土壤、品种和气候等条件有关，没有完整的田间记录是很难鉴定的，最好是就地观察、调查和分析，不能单纯依靠室内检查。

2. 植物病原物的鉴定

(1)菌物的鉴定

菌物是最常见的植物病原物类群，引起绝大多数植物病害。菌物(真菌)引起的植物病害除了有各种病状外，大多还会产生病征，如粉、锈、霉、点、蕈菌等。根据病征及病原物的分类特征，可进一步区分菌物的类群。基于症状观察结果，对植物病害标本进行显微镜观察是非常重要的。菌物(真菌)病害标本或培养物，应根据材料的不同而选用不同的镜检方法。

①菌丝体和子实体的挑取检视：标本或培养基上的菌丝体或子实体，一般可直接挑取少许，制成临时显微玻片进行镜检。

②叶面菌物(真菌)的粘贴检视：可用透明胶、醋酸纤维素、火棉胶或其他胶黏剂直接胶黏标本表面的真菌，如白粉病、霜霉病、锈病等。

③徒手切片检视：该方法在教学和科研上用处很大，一般是切取植物病害标本病部的小块组织，将其夹在通草中固定，用刀片直接进行切片。此法不需要特殊设备(切片机)，操作方便，是最常用的真菌检视方法。通过反复练习，切取的植物组织可以非常薄，在显微镜下既能观察到发病植物细胞的形态，也能观察到病原物的子实体结构，对病原物的鉴定有非常重要的作用。

④组织整体透明检视：组织透明后的整体检查，不但可以看清表面的病菌，还可以观察组织内部的病菌。使组织透明的方法很多，最常用的是水合氯醛：将小块发病组织在等量的乙醇和冰醋酸的混合液中固定 24 h，然后浸泡在饱和的水合氯醛溶液中，待组织透明后取出用水洗净，经苯胺蓝水溶液染色，用甘油浮载检视。

⑤真菌的玻片培养检视：采用培养基培养的真菌，直接挑取菌落表面物质检视往往会破坏其固有结构，故玻片培养是很好的检视方法。

(2)细菌的鉴定

细菌性植物病害在受害部位的维管束或薄壁细胞组织中一般都有大量的菌体，其徒手切片在显微镜下常能观察到“喷菌”现象。对植物病原细菌的具体鉴定，一般都要用分离培养法以获得纯培养物，然后对纯菌株的生理生化指标和致病性进行测定，从而将其鉴定。

(3)病毒的鉴定

病毒的鉴定首先根据其症状特点进行，植物病毒病的病状一般为花叶、变色、坏死、矮缩、矮化和畸形等。通过常规的光学显微镜检视，可以观察到部分病毒在发病组织部位形成的内含体，根据内含体的形态特点可以对病毒的种类进行初步判断。病毒的精准鉴定需要借助电子显微镜和分子生物学技术，本实习不做要求。

(4)线虫的鉴定

植物受到线虫危害后，可形成各种症状，如根结、胞囊、叶斑、坏死或萎蔫等。首先，基于所学的理论知识初步判断调查的植物病害是否由线虫引起；其次，可以进行线虫的分离和挑取在显微镜下进行线虫的形态观察。对一些虫体较大的线虫，可在解剖镜下直接挑取进行显微镜观察，或者直接制作徒手切片在显微镜下观察虫体，如根结线虫病、胞囊线虫病等。

3. 显微成像及绘图

(1)显微摄影

通过显微成像系统可对显微镜下观察到的目标物进行拍照和保存，对病原物的鉴定和资料交流至关重要。其步骤如下：

①玻片制作：把需要进行显微摄影的病原物制成临时玻片。

②镜检：在显微镜下观察，找到需要拍照的结构。

③设置参数：打开显微成像系统，设置相关参数，务必注意设定放大倍数要与物镜当前所用倍数一致，还可以设置比例尺的放置位置、方式和颜色等。

④拍照：设置好参数后，调节显微镜的细准焦螺旋，使病原物结构在计算机屏幕上尽可能清晰地显示，点击拍照后将照片保存。

(2)显微绘图

在使用显微镜时，采用徒手描绘显微镜下观察到的病原物的结构是非常必要的。当然，可以采用显微描绘仪进行精准绘图。在本实习中，要求学生进行徒手绘图。

4. 注意事项

(1)以寄主植物分类地位为主、结合病原物形态特征进行鉴定

此类病害包括锈病、白粉病和霜霉病等。其中，植物锈菌因其具有专性寄生和寄生专化性，确定寄主植物的分类地位对锈菌的准确鉴定尤为重要。白粉菌和霜霉菌对寄主植物也有一定的选择性，寄主种类鉴定也是其病原菌鉴定的前提。

(2)以病原物形态特征为主、结合寄主植物分类地位进行鉴定

此类病害如叶斑病、溃疡病、枝枯病及萎蔫，其病原菌多为子囊菌及其无性型真菌，为兼性寄生菌或兼性腐生菌，涉及种类多、形态复杂多样。因此，可依据病征类型采取合适的显微制片方法。通过真菌形态学观察，可初步确定病原真菌的分类地位，再根据寄主植物种类、查阅相关资料做出准确鉴定。在此类病害鉴定过程中，应注意以下问题，以免误判。

①同原不同阶段：由于此类病害多为子囊菌，且以其无性型在生长季节多见。同一种病原真菌的有性和无性阶段可同时表现在同一病害病状中。要全面检查多份病害标本，对病原菌的无性型和有性型做出完全的鉴定

②同症异原：此种情况较为复杂。除了上述同原不同阶段的情况外，有时在相同的病斑上，也会观察到形态和分类地位完全不同的两种病原真菌。此外，子囊菌多为兼性寄生菌，在其所致坏死病斑上，在后期往往会长出腐生真菌，可能会掩盖了真正的病原菌，因

此，发病初期病害标本的诊断至关重要。

③同原异症：此种情况较少。常见的如苹果褐斑病（*Marssonina mali*）在叶片上会出现针芒型、同心轮纹型和混合型三种病斑；柑橘炭疽病（*Colletotrichum gloeosporioides*）在叶片和枝干上病状也有所差异。

④新病害：应遵循科赫法则进行准确鉴定。

【作业和思考题】

1. 将在实习中进行的植物病害标本鉴定情况填入表3-4。

表 3-4　植物病害标本鉴定表

标本号	寄主植物	病害名称	病原物学名	采集地	采集人	采集时间

2. 每人提交20张优质的显微摄影图片，并在相应的照片上进行标注病害名称、病原物学名、放大倍数和鉴定人。

3. 植物病原物鉴定的经典方法和分子生物学方法有哪些？各有何特点？

4. 谈谈植物病原物鉴定的重要性。

5. 如何有效区分侵染性和非侵染性植物病害。

PRACTICE 3　Diagnosis of Plant Diseases and Pathogen Identification

【Introduction】

Plant disease diagnosis is the application of the basic knowledge and theory of plant pathology to the practice of agricultural production, which is the premise for the control of plant diseases. In the face of a plant disease, we should first determine whether it is infectious or noninfectious. If it is an infectious plant disease, the following operations including comprehensive observation of symptoms, microsection preparation and microscopic examination, and isolation and cultivation of the pathogen should be carried out to further identify the specific pathogen.

【Purpose and Requirements】

1. Understand the occurrence characteristics of different types of plant diseases, and master the general methods for plant disease diagnosis and pathogen identification.

2. Master the specific steps of plant disease diagnosis and the slide preparation methods, and put forward the reasonable control plan of plant diseases.

【Materials and Appliances】

1. Materials

Various fresh plant disease specimens.

2. Tools

Microscopes, microscopic imaging system, magnifying glasses, slides, coverslips, blades, ricepaperplant piths, tweezers, alcohol, distilled water, absorbent papers, lens tissues, dissecting needles, pencils, drawing papers, identification record books, identification labels, etc.

【Contents and Methods】

1. Diagnosis of Plant Disease

Plant diseases can usually be initially diagnosed based on the characteristics of their symptoms. Firstly, the plant disease can be preliminarily determined whether it is infectious or noninfectious according to the occurrence characteristics in the field. Secondly, the symptom characteristics of the collected plant disease specimens should be carefully observed and recorded in detail in the identification record notebook. For the infectious diseases with obvious symptom characteristics, it is necessary to further determine the specific type of pathogens, such as fungi, bacteria, viruses or nematodes, according to the symptom characteristics of plant diseases caused by various types of pathogens. Pay attention to the following precautions when observing symptoms:

①When observing symptoms, attention should be paid to the effect of the disease on the whole plant before inspecting the diseased site.

②When observing spot disease, attention should be paid to the number, size, color, arrangement, and whether there are wheel lines of spots. When observing the rotting disease, we should pay attention to the color, odor, and structure(such as soft rot, dry rot, and wet rot) of the rotting tissue and whether there is insect injury, etc.

③Be sure to pay attention to whether there are signs on diseased sites, which play a very important role in the identification of pathogens. Some plant diseases caused by fungi can form the signs of powder, rust, mould, particles, and mushroom on the diseased sites. Some bacterial plant diseases may produce bacterial ooze. There are no signs for the plant diseases caused by viruses and *Phytoplasma*.

④ Many noninfectious diseases may be encountered during the identification process. Sometimes, they have the similar symptoms with viral diseases, nematode diseases, or some fungal diseases, and even some fungi can be detected on the diseased sites. Of course, these fungi are generally saprophytes growing on the necrotic tissues after the plant is damaged by noninfectious diseases.

⑤Noninfectious diseases commonly occur in the field with not obvious infection center, which are related with terrain, soil, varieties, climate, and other conditions. They are difficult to identify without complete field records. It is better to observe, investigate, and analyze diseases *in situ* rather than simply rely on indoor detection.

2. Pathogen Identification

(1) Identification of Fungi

Fungi are the most common group of plant pathogens, causing most plant diseases. In addition to various morbidities, most of the plant diseases caused by fungi also produce signs, such as powdery, rust, mildew, particles, and mushrooms, etc. According to the signs and classification characteristics of pathogens, the groups of fungi can be further distinguished. Based on the observation of symptoms, it is very important to conduct the microscopy observation of plant disease specimens. For fungal disease specimens or cultures, different microscopy examination methods should be selected according to different materials.

①Picking mycelia and fruiting bodies for observation: Mycelia or fruiting bodies on specimens or culture media can be picked up directly and made into temporary slides for microscopy observation.

②Pasting foliar fungi for observation: Transparent tape, cellulose acetate, collodion, or other adhesives can be used to directly paste the fungi on the surface of specimens, such as powdery mildew, downy mildew, rust, etc.

③Hand-making section for observation: This method is very useful in teaching and scientific research. It is generally used to cut small pieces of tissue from the disease site of plant disease specimens, clip them in ricepaperplant pith and fix them, and slice them directly with a blade. This method does not require special equipments (slicer), and is easy to operate. Through repeated practice, the cut plant tissue can be very thin, and the morphology of the diseased plant cells and the fruiting body structure of the pathogen can be observed under the microscope. This method plays a very important role in the identification of pathogens.

④ Transparency of whole tissue for observation: The overall examination after the transparency of the tissue can not only see the pathogens on the surface, but also observe the pathogens inside the tissue. There are many ways to make tissue transparent. The most commonly used method is using chloral hydrate: Fix small pieces of diseased tissue in the mixture of alcohol and glacial acetic acid for 24 h, and then soak them in saturated chloral hydrate solution, wash the tissue with water when it is transparent, stain it with aniline blue aqueous solution, and examine it with glycerin floating.

⑤Slide culture of fungi for observation: The intrinsic structures of the fungi cultured on the medium will be damaged if the materials of colony surface are directly picked up for observation. Hence, the glass slide culture is a good method for observation.

(2) Identification of Bacteria

For bacterial plant diseases, there are usually a large number of bacterial individuals in the vascular bundle or parenchyma cell tissue of the affected site, and the phenomenon of "bacteria exudation" can be observed in the handworked section under the microscope. For specific identification of plant pathogenic bacteria, isolation and culture method is generally used to obtain the pure culture, and then identify them by determining the physiological, biochemical characteristics and pathogenicity of the pure strains.

(3) Identification of Viruses

The identification of virus can first be conducted according to the characteristics of symptoms.

The common morbidities of plant viral diseases include mosaic, discoloration, necrosis, dwarf, stunt and malformation. The inclusion bodies of some viruses in the diseased sites can be observed by routine light microscope examination, and the types of viruses can be preliminarily identified according to the morphological characteristics of the inclusion bodies. The accurate identification of viruses requires the use of electron microscopy and molecular biology techniques, which is not required in this internship.

(4) Identification of Nematodes

After being damaged by nematodes, plants can show various symptoms, such as root knots, cysts, leaf spot, necrosis, and wilting, etc. Firstly, we can judge whether the investigated disease is caused by nematodes or not based on the theoretical knowledge we have learned, and then we can separate and pick out nematodes for microscopic morphological observation. For some large nematodes, they can be picked up directly under the anatomical microscope for observation, or observed under the microscope using hand-making section, such as root knot nematode and cystic nematode.

3. Microscopic Imaging and Drawing

(1) Microphotography

The microscopic imaging system can photograph and preserve the objects observed under the microscope, which is very important for pathogen identification and data exchange. The steps are as follows:

①Slide preparation: Make temporary slides of the pathogens to be microphotographed.

②Microscopy: Observe and find the structures to be photographed under a microscope.

③Set parameters: Turn on the microscopic imaging system and set the relevant parameters. Be sure that the magnification ratio should be consistent with the current multiple of the objective lens. You can also set where, how, and color of the scale, etc.

④Photograph: After setting the parameters, adjust the fine-focus screw of the microscope to display the pathogen structure as clear as possible on the computer screen. Click to take a picture and save the picture.

(2) Microscopic Drawing

When using a microscope, it is very necessary to draw the structure of the pathogen observed under the microscope by hand. Of course, the accurate drawing can be done with a microprofiler. In this practice, students are required to draw by hand.

4. Precautions

(1) Identification Mainly Based on Taxonomic Status of Host Plants Combing with Morphological Characteristics of Pathogen

Such diseases include rust, powdery mildew, and downy mildew. The taxonomic status of host plants is particularly important for the accurate identification of rust due to its obligate parasitism and parasitic specificity. The host species identification is also the premise for the identification of powdery mildew and downy mildew, as these pathogens are also selective to host plants.

(2) Identification Mainly Based on Morphological Characteristics of Pathogen Combing with Taxonomic Status of Host Plants

Such diseases include leaf spot, canker, branch blight, and wilt, the pathogens of which are mostly the fungi in Ascomycota and their anamorph. They are facultative parasites or facultative saprophytes, involving a variety of complex and diverse forms. Therefore, according to the type of symptoms, appropriate microscopic slide method should be adopted. The taxonomic status of pathogenic fungus can be preliminarily determined by the morphological observation. Then the accurate identification can be carried out by referring to relevant information and the species of host plants. In the process of identifying such diseases, attention should be paid to the following problems to avoid misjudgment.

① Different stages of the same pathogen: Since these pathogens mainly belong to Ascomycota, the anamorph of them is more common in the growing season of host plants. The teleomorph and anamorph of the same pathogenic fungus can be simultaneously manifested in the same morbidity. Multiple disease specimens should be thoroughly examined and the teleomorph and anamorph of the pathogen should be completely identified.

②Same symptom with different pathogens: This situation is more complex. In addition to the abovementioned different stages of the same pathogen, sometimes two pathogenic fungi with completely different morphologies and taxonomic statuses can also be observed on the same spot. Besides, the fungi in Ascomycota are mainly facultative parasites, and some saprophytic fungi often grow on necrosis spot in the late stage, which might cover up the true pathogen. Therefore, the diagnosis of disease specimens in the early stage of disease is extremely important.

③Different symptom with same pathogens: It's rare. Such as apple brown spot (*Marssonina mali*), there are three types of spots on the leaves, including radial type, concentric ring, and mixed type, while for citrus anthracnose (*Colletotrichum gloeosporioides*), the symptoms on leaf and branch are also different.

④New diseases: They should be identified according to Koch's postulate.

【Assignments and Questions】

1. Fill the identification results of plant disease specimens in this practice into Table 3-4.

Table 3-4 Identification Table of Plant Disease Specimens

Specimen No.	Host plant	Disease name	Pathogen name	Collection location	Collector	Collection time

2. Submit 20 high-quality photomicrographic pictures for each student, and mark the disease

name, pathogen name, magnification and appraiser on the corresponding photos.

3. What are the classic methods and molecular biological methods for the identification of plant pathogens? What are their characteristics?

4. Talk about the importance of plant pathogen identification.

5. How to effectively distinguish between infectious and noninfectiuos plant diseases?

参考文献
References

阿格里斯 G N. 植物病理学 [M]. 5 版. 沈崇尧，译. 北京：中国农业大学出版社，2009.

毕朝位，陈国康. 普通植物病理学实验实习指导 [M]. 重庆：西南师范大学出版社，2017.

曹支敏，李振岐. 秦岭锈菌 [M]. 北京：中国林业出版社，1999.

程丽娟，薛泉宏. 微生物学实验技术 [M]. 2 版. 北京：科学出版社，2012.

方中达. 植病研究方法 [M]. 3 版. 北京：中国农业大学出版社，1998.

候明生，黎丽. 农业植物病理学实验实习指导 [M]. 北京：科学出版社，2014.

李玲. 植物生理学模块实验指导 [M]. 北京：科学出版社，2009.

李培琴，费昭雪，闫金娇. 普通植物病理学实践教程 [M]. 咸阳：西北农林科技大学出版社，2021.

李培琴，明洁，张舒遥，等. 黄帝陵侧柏叶枯病的病原菌鉴定及防治药剂筛选 [J]. 西北农林科技大学学报(自然科学版)，2021，49(1)：74-84.

刘维志. 植物病原线虫学 [M]. 北京：中国农业出版社，2000.

陆家云. 植物病原真菌学 [M]. 2 版. 北京：中国农业出版社，2001.

阮钊，丁俊园，唐光辉，等. 花椒根腐病的病原鉴定与防治药剂筛选[J]. 植物病理学报，2022，52(4)：630-637.

田呈明. 森林菌物与病害认知 [M]. 北京：中国林业出版社，2018.

王学奎. 植物生理生化实验原理和技术 [M]. 2 版. 北京：高等教育出版社，2006.

谢联辉，林齐英. 植物病毒学 [M]. 2 版. 北京：中国农业出版社，2004.

许文耀. 普通植物病理学实验指导 [M]. 北京：科学出版社，2006.

许志刚. 普通植物病理学 [M]. 3 版，北京：中国农业出版社，2003.

许志刚. 普通植物病理学实验实习指导[M]. 2 版，北京：高等教育出版社，2008.

许志刚. 普通植物病理学 [M]. 4 版. 北京：高等教育出版社，2009.

张铉哲，冉隆贤，李永刚，等. 植物病理学研究技术 [M]. 北京：北京大学出版社，2015.

张志良，李小芳. 植物生理学实验指导 [M]. 5 版. 北京：高等教育出版社，2016.

赵斌，林会，何绍江. 微生物学实验 [M]. 2 版. 北京：科学出版社，2014.

郑用琏. 基础分子生物学 [M]. 3 版. 北京：高等教育出版社，2018.

周长林. 微生物学实验与指导 [M]. 3 版. 北京：中国医药科技出版社，2015.

AGRIOS G N. Plant pathology [M]. 5th Edition. Burlington：Academic Press，2005.

LI P Q，RUAN Z，FEI Z X，et al. Integrated transcriptome and metabolome analysis revealed that flavonoid biosynthesis may dominate the resistance of *Zanthoxylum bungeanum* against stem canker [J]. Journal of Agricultural and Food Chemistry，2021，69(22)：6360-6378.

MOSCATELLO S，PROIETTI S，BUONAURIO R，et al. Peach leaf curl disease shifts sugar metabolism in severely infected leaves from source to sink [J]. Plant Physiology and Biochemistry，2017，112：9-18.

ZHOU X，O'DONNELl K，AOKI T，et al. Two novel *Fusarium* species that cause canker disease of prickly ash (*Zanthoxylum bungeanum*) in northern China form a novel clade with *Fusarium torreyae* [J]. Mycologia，2016，108(4)：668-681.

附录I　植物病害的症状类型

1. 植物病害的病状类型

病状是指植物患病后其本身所表现出来的不正常状态。病状类型很多，根据组织病变的性质，一般分为三大类：坏死性病状、促进性病状和抑制性病状。坏死性病状常以植物细胞和组织死亡为特征，表现为坏死斑、焦枯、腐烂等；促进性病状是指植物细胞和组织受到病原物的刺激而膨大或增生；抑制性病状是指植物的生长发育部分或全部受到抑制。常见的植物病状类型包括变色、坏死、萎蔫、腐烂和畸形五大类。

(1) 变色

变色是指发病植物色泽发生改变，大多数发生在叶片上，由于叶绿素的形成受阻或破坏而发生不同程度的褪绿或出现其他色素(附图 1-1)。有时，果实、种子和花瓣也会出现各种变色。变色的主要类型如下：

①褪绿或黄化：褪绿是指整片叶子或叶片部分均匀地变色。由于叶绿素含量在整个叶片中的含量均匀地减少到一定程度，则出现黄化。

②花叶与斑驳：叶片不均匀的变色称为花叶或斑驳。花叶是由形状不规则的深绿、浅绿、黄绿或黄色的变色斑相嵌形成的杂色，不同变色部分轮廓清晰，如苹果花叶病、牡丹花叶病、烟草花叶病等；如果变色部分的轮廓不清晰，称作斑驳。

③碎锦：发生在花瓣上的变色称为碎锦，其能使花瓣色彩更加绚丽多彩。

④脉明：是指叶片的主脉和支脉因褪色而变为半透明状。

附图 1-1　植物病害病状——变色

(2) 坏死

坏死是指发病植物局部或大面积细胞和组织的死亡，常随病原和发病部位的不同而表现不同的坏死类型(附图 1-2)。主要类型如下：

褐斑

黑斑

白斑

条斑

圆斑

角斑

叶枯

炭疽

疮痂

溃疡

溃疡

猝倒

枝枯

附图 1-2　植物病害病状——坏死

①斑点：是指患病植物组织局部坏死而形成的病斑，一般有明显的边缘，可因形状、大小和颜色的不同而分为不同类型，如褐斑、黑斑、灰斑、紫斑、白斑、条斑、圆斑、角斑、环斑、大斑、小斑等。

②炭疽：是斑点的一种，常指由炭疽菌所致的斑点病状。病斑中常可见散生或轮生的黑色小颗粒，为炭疽菌的分生孢子盘。

③疮痂：与斑点近似，但在病斑上有增生的木栓层，使其表面粗糙或是病斑枯死后，因生长不平衡使发生龟裂，如柑橘疮痂病、梨黑星病、泡桐黑痘病等。

④溃疡：是指树木枝干上的大片皮层组织坏死，常使其木质部外露，病部周围由于组织愈伤作用而稍微隆起，常呈同心环状。多见于木本植物的枝干，如板栗疫病和杨树溃疡病。

⑤猝倒和立枯：发生于各种植物的幼苗期。由于幼苗茎基部组织受到侵染而坏死，有时引起植物突然倒伏形成猝倒，有时虽然坏死但不倒伏的称为立枯。

⑥焦枯：是指植物的叶、花、穗、芽等器官局部或全部坏死，又常称为疫病或瘟病。它是由病斑的发展或联合而造成植物的整体性死亡，枯死面积通常占植物组织面积的1/3以上。

(3)萎蔫

萎蔫是指植物整株或局部因脱水而枝叶萎垂的现象。典型的萎蔫症状是植物根茎的维管束组织受到破坏而发生的凋萎现象，而根茎的皮层组织还是完好的。病理性萎蔫是由于植物输导组织受到病原物的毒害或阻塞所致，与生理性缺水萎蔫不同，不能因供水而恢复。萎蔫病害常无外表的病征。根据症状的不同，萎蔫又可分为枯萎、黄萎和青枯等类型(附图1-3)。

枯萎　　黄萎　　青枯

附图1-3　植物病害病状——萎蔫

(4)腐烂

腐烂指植物组织较大面积的分解和破坏，腐烂与坏死有时不易区别。植物的根、茎、叶、花、果都可发生腐烂，幼嫩或多肉的组织则更容易发生；根据组织分解程度的不同，又可分为湿腐、软腐和干腐，且伴随各种颜色的变化，如褐腐、白腐、黑腐(附图1-4)。木本植物常因蕈菌危害造成的枝干腐烂又称为腐朽。

根腐

花腐

果腐

干腐

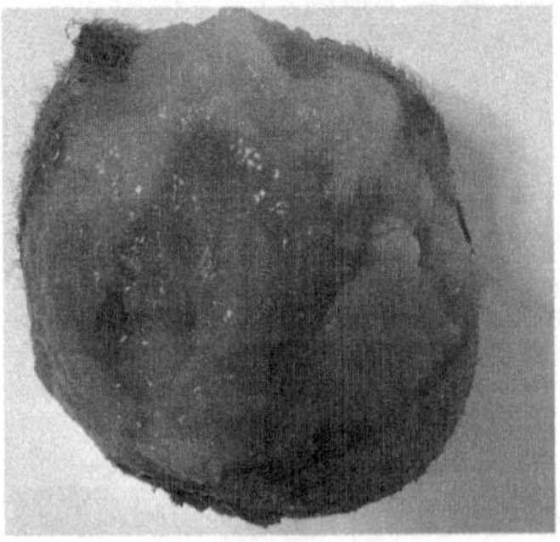
软腐

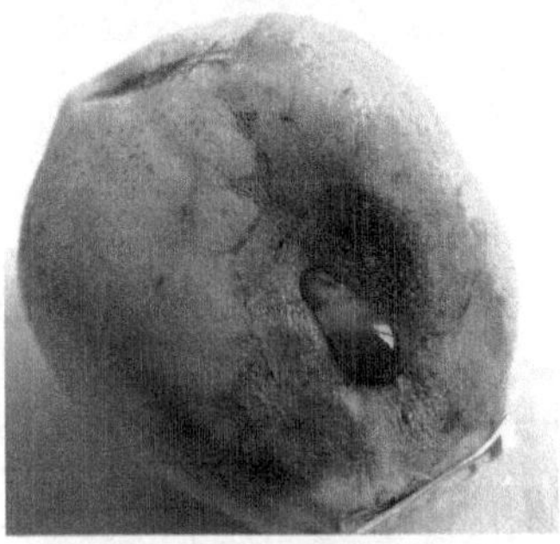
湿腐

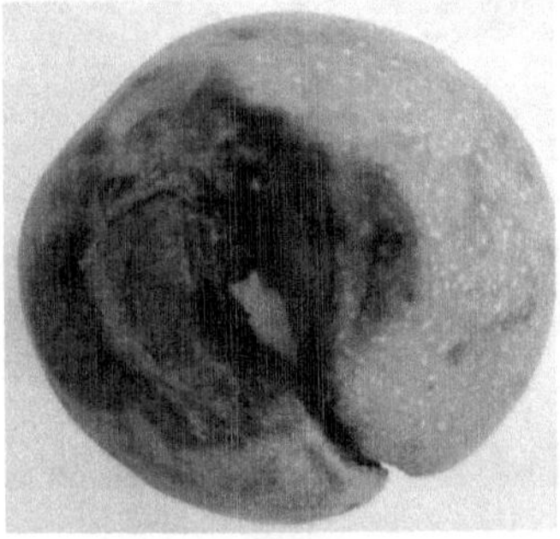
黑腐

褐腐

白腐

腐朽

附图 1-4　植物病害病状——腐烂

(5)畸形

畸形是指植物受害部位的细胞分裂和生长发生促进性或抑制性的病变，植物整体或局部的形态异常(附图 1-5)。畸形通常可分为增大、增生、减生和变态 4 种类型。

矮化

缩叶

丛枝

干瘤

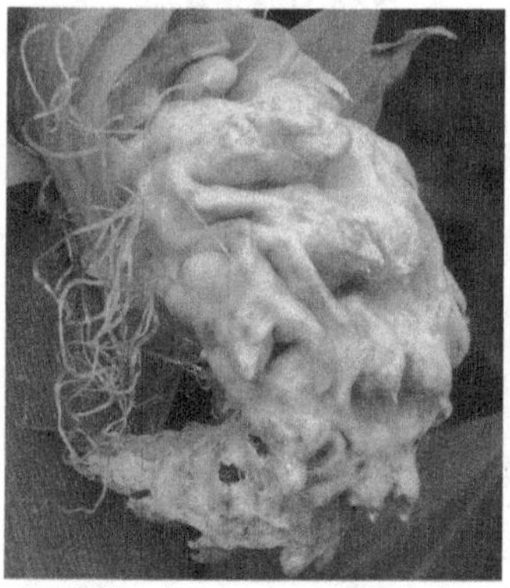
粉瘤

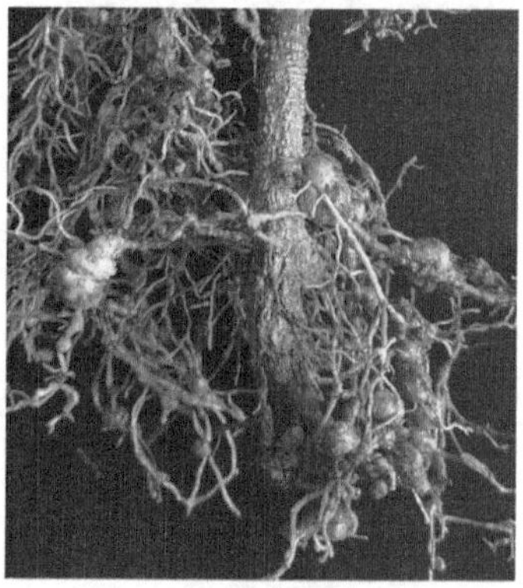
根结

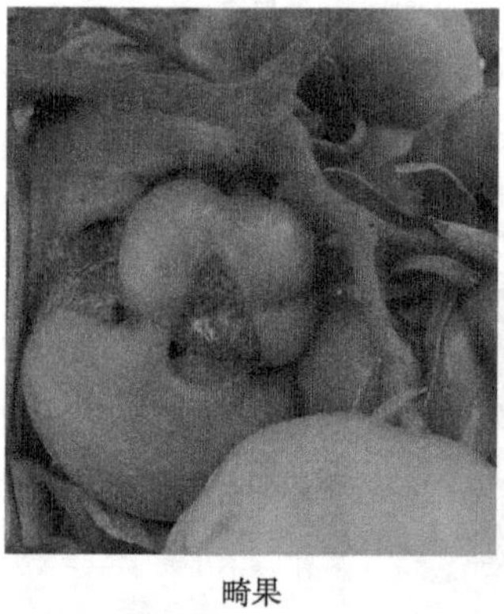
畸果

花变叶

带化

附图 1-5　植物病害病状——畸形

①矮缩：指植物节间缩短或停止生长，病株比健株矮小得多，如水稻矮缩病、小麦黄矮病、玉米矮化病等。

②丛生：指植物的枝条或侧根异常增多，导致丛枝或丛根，如枣疯病、泡桐丛枝病、苹果发根病、竹丛枝病等。

③瘤肿：指病部的细胞或组织因病原物的刺激而增大或增生，表现出瘤肿，如葡萄根癌病、白菜根肿病、樱花冠瘿病、玉米瘤黑粉病、花椒根结线虫病、大豆胞囊线虫病等。

④缩叶：指叶片卷曲和皱缩，有时病叶变厚，变硬，严重时呈卷筒，如桃缩叶病、马铃薯卷叶病、蚕豆黄化卷叶病等。

⑤蕨叶：指叶片变成丝状、线状或蕨叶状，如番茄蕨叶病、辣椒蕨叶病等。

⑥畸果：指果实变形。

⑦小果：指发病果实比正常果实瘦小。

⑧花变叶：指发病植物的花器变成叶片结构，使植物不能正常开花结果，也称为花器返祖，如月季绿萼病、芍药绿萼病等。

⑨带化：指发病植物的枝干变为扁平状，如苹果枝干带化病、黄瓜茎带化病、柽柳枝干带化病等。

2. 植物病害的病征类型

病征是由病原菌物(偶为细菌)在发病植物组织表面产生的营养体、繁殖体或病害产物而表现出的特征。植物病原细菌只表现外形相似的病征，而不同种类的病原菌物之间有较明显的形态差异，其病征的特点也不相同。常见的病征包括粉、霉、点状物、菌核、蕈菌、菌脓等。

(1)粉

有些植物叶片、枝条或果实被病菌侵染后，在被害组织的表面会产生一层粉状物，因其形状和颜色的差异可进一步分为黄锈、白粉和黑粉(附图 1-6)。

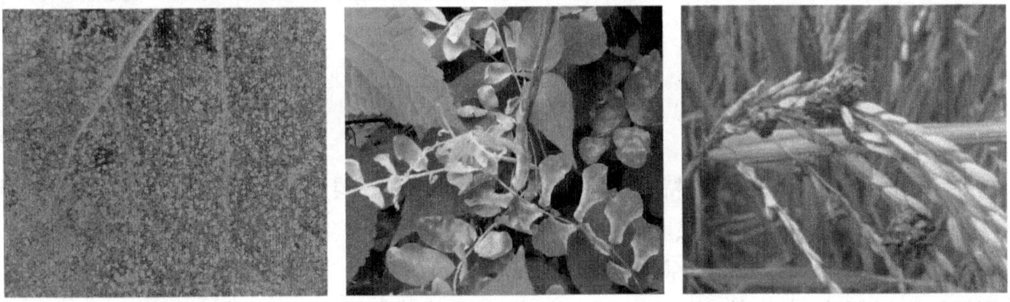

黄锈　　白粉　　黑粉

附图 1-6　植物病害病征——粉

①黄锈：指颜色为鲜黄、橘黄、褐色或黑褐色的粉状物，常为各种植物锈病的特有病征。

②白粉：指颜色为白色的粉状物，常为白粉病的特有病征。初期为白色，后期转为褐

色，并在粉层中产生很多黄褐色、最后变为黑色的球形颗粒(闭囊壳)。

③黑粉：指颜色为黑色的粉状物，常产生于发病植物的穗部、籽粒内外、叶或茎内部。黑粉数量很大，特别显著，是黑粉菌孢子形成的特征，为各种植物黑粉病专有特征。

(2)霉

该病征是由各种菌物的菌丝、分生孢子、分生孢子梗、孢子囊和孢囊梗等构成。霉层的颜色、形状、结构和疏密等特点的差异，可标志不同的病原物(附图 1-7)。

附图 1-7　植物病害病征——霉

①霜霉：为植物霜霉病的特有病征。在结构上，该霉层由病原菌的孢囊梗和孢子囊构成，颜色多为白色，也有灰色、紫色或黑色。

②毛霉：为一些腐烂性病害的病征，霉层丰厚，初期为白色，后期转为黑白相间或表面密生一层黑色球状体(孢子囊)，多为毛霉菌或根霉菌所致的果实和块茎腐烂的病征。

③青霉：发生于多种果实、块根、块茎的腐烂病体上，为青绿色，多为青霉菌侵染所表现的病征。

④灰霉：可危害多种植物的果实和花器，主要由灰霉属真菌引起，表现为灰色霉层，该霉层由病菌的菌丝体、分生孢子和分生孢子梗构成。

⑤黑霉：该病征常伴随着多种病状而产生，主要是坏死性病状，霉层特征差异明显，

或疏松，或致密，一般较薄。主要为无性型真菌所致病害的病征。

(3)点状物

该病征是指在植物发病组织部位的表面上产生的小颗粒状物，大多为病原物的子实体，如子座、闭囊壳、子囊壳、分生孢子器、分生孢子盘等。不同病害产生的点状物的形状、大小、颜色、突起程度、密度和数量均不尽相同。点状物的常见颜色有黑色、蓝紫色、橘红色等(附图 1-8)。

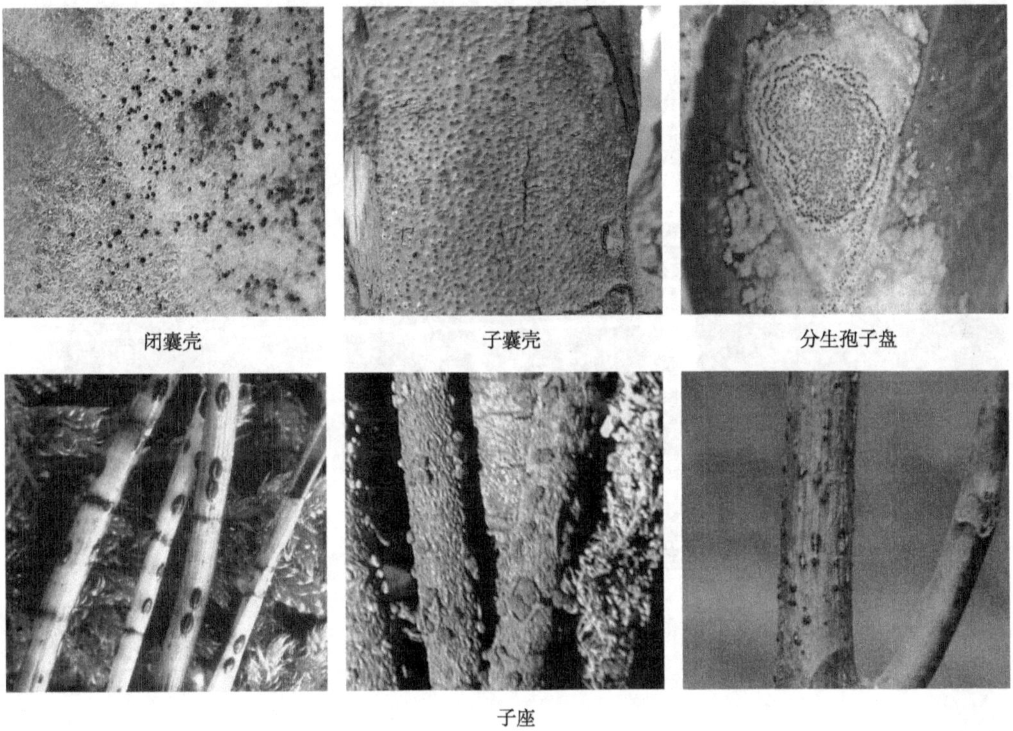

附图 1-8　植物病害病征——点状物

(4)菌核

菌核是指由真菌菌丝交织形成的一种休眠结构，形状大小差别很大，多数为黑色的，少数棕色，常伴随着整株或局部的腐烂或坏死而形成，产生于植物病组织表面或茎秆内部髓腔中，此类病害多称为菌核病(附图 1-9)。

(5)蕈菌

蕈菌通常指生长于树木枝干或腐朽木材上的形态各异的高等担子菌的子实体，如菌伞、菌苔、木蹄等，使被害树干木质部腐朽(附图 1-9)。

(6)菌脓

菌脓为多数细菌性植物病害的特有病征，可在病部观察到菌脓，其中含有大量的细菌菌体，常为白色、绿色或黄色，干涸时形成菌胶粒或菌膜(附图 1-9)。

(7)其他

锈菌可在发病植物部位形成毛柱状、舌状或杯状等不同形状的锈孢子器，或形成瘤状的冬孢子菌瘿；某些真菌病害还可以形成根状菌索结构或产生流胶现象(附图 1-9)。

菌核

蕈菌

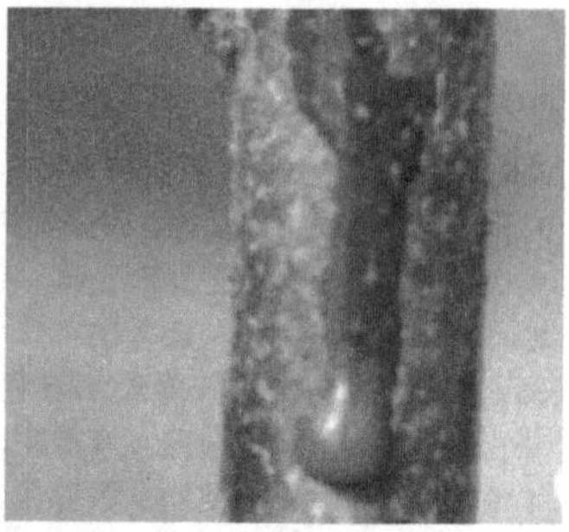

菌脓（引自许志刚，2009）

锈子器

菌瘿

流胶

附图 1-9　植物病害的其他病征

Appendix I　Symptom Types of Plant Diseases

1. Morbidity Types of Plant Diseases

Morbidity of plant disease refers to the abnormal state of the plant itself after it is sick. There are many types of morbidities, which are generally divided into three categories according to the properties of tissue changes, including necrotic morbidity, promoted morbidity and suppressive morbidity. Necrotic morbidity is characterized by the death of plant cells and tissues, which is shown by necrotic spots, scorch, or decay, etc. Promoted morbidity refers to the swelling or proliferation of plant cells and tissues stimulated by pathogens. Suppressive morbidity means that the plant growth and development are partially or totally inhibited. The common mobidity of plant diseases includes five types, i. e. discoloration, necrosis, wilt, decay and malformation.

(1) Discoloration

It refers to the changes of color colour and lustre of diseased plant, mostly occurring on the leaf. As the formation of chlorophyll is blocked or destroyed, different degrees of chlorosis occurs or other pigments are shown (Appended figure 1-1). Sometimes, fruits, seeds, and petals also appear a variety of discoloration. The main types of discoloration are as follows:

Appended figure 1-1 Morbidities of Plant Diseases——Discoloration

①Chlorosis or yellowing: It refers to the uniform discoloration of the whole or part of leaf. When the chlorophyll amount in the leaf is uniformly reduced to a certain extent, yellowing occurs.

②Mosaic and mottle: It refers to the uneven discoloration of leaves. Mosaic is a kind of variegated colors, which is formed by dark green, light green, yellow-green or yellow variegated spots with irregular shape. The outlines of different discolored parts are clear, such as apple mosaic, poeny mosaic and tobacco mosaic, etc. If the outline of the discolored part is not clear, it is called mottle.

③Color breaking: It refers to the discoloration on the petals, which can make the petals more colorful.

④Vein clearing: It refers to that the main veins and branches of the leaf become translucent due to discoloration.

(2) Necrosis

Necrosis refers to the local or large area of the death of cells and tissues in the diseased plant, which often shows different types of necrosis due to the differences of pathogen and infection tissues (Appended figure 1-2). The main types are as follows.

①Spots: It often refers to the lesions formed by local necrosis of diseased plant tissue, generally having obvious edges, which can be classified as different types due to the differences of shape, size, and color, such as brown spots, black spots, gray spots, purple spots, white

Brown spot

Black spot

White spot

Stripe spot

Round spot

Irregular spot

Leaf blight

Anthracnose

Scab

Canker

Canker

Damping off

Branch blight

Appended figure 1-2　Morbidities of Plant Diseases——Necrosis

spots, stripes, round spots, angular spots, ring spot, large spots, and small spots, etc.

②Anthracnose: It is a kind of spot, which often refers to the spot caused by the fungi in the genera of *Collectotrichum*. Some black particles can usually observed on the spots showing a scattered or cyclic arrangement, which are the acervuli of *Collectotrichum* spp.

③Scab: It is similar to the spot, but there is a proliferating cork layer on the spot, making the surface rough or crack after the spot dies as the unbalanced growth, such as citrus scab, pear black scab, paulownia black pox, etc.

④Canker: It refers to a large area of necrotic cortical tissue on the stem of a tree, which often results in exposed xylem. The surrounding area of the diseased area is slightly raised due to callus, often in a concentric ring. It is common in woody plant stems, such as chestnut blight and poplar canker.

⑤Damping off and sheath blight: It occurs in the seedling period of various plants. Due to the necrosis in the infected basal tissue of seedling stem, the plant sometimes collapses suddenly and forms a damping-off, while it is called sheath blight when the whole plant is necrotic but not damps off.

⑥Blight: It refers to the partial or complete necrosis of leaf, flower, ear, bud, and other organs, which is also called as blight or blast. It is the overall death caused by the development or combination of spots, with the dead area accounting for more than 1/3 of the plant tissue area.

(3) Wilt

It refers to the phenomenon that the wilt of branches and leaves caused by dehydration of whole or part of the plant. The typical symptom of wilt is the withering of the vascular tissues of root and stem, while the cortical tissue is still intact. Pathological wilt is caused by the damage or obstruction of plant condcuting tissues by pathogens, which is different from physiological water shortage wilt and cannot be recovered by water supply. There are usually no outward signs for wilt diseases. According to the different symptoms, wilt can be divided into wither, yellow wilt, and green wilt, etc(Appended figure 1-3).

Wither　　Yellow wilt　　Green wilt

Appended figure 1-3　Morbidities of Plant Diseases——Wilt

(4) Rot

Rot refers to the breakdown and damage of the larger area of plant tissue, which is sometimes not easy to distinguish from necrosis. Rot can occur on root, stem, leaf, flower and fruit, especially on young or fleshy tissue. It can be divided into wet rot, soft rot, and dry rot according to the decomposition degree of tissues, which is accompanied with all sorts of color change, such as brown rot, white rot, black rot, etc (Appended figure 1-4). The stem rot of woody plant caused by mushroom fungi is called as decay.

(5) Malformation

It refers to the promoted or inhibited morbidity of cell division and growth in injured tissues of plant, which then results in the morphological abnormality of the whole or part of plant (Appended figure 1-5). It can often be divided into hypertrophy, hyperplasia, hypoplasia and metamorphosis.

①Dwarf: It refers to the shortening or atrophy of plant internode. The diseased plant is much shorter than the healthy plant, such as rice dwarf, wheat yellow dwarf, and maize dwarf, etc.

②Witches'broom: It refers to the abnormal increased branches or lateral roots of a plant, resulting in the arbuscules of branches or roots, such as jujube withes broom, paulownia withes broom, apple hairy root, bamboo withes broom, etc.

③Tumor: It refers to the enlargement or hyperplasia of cells or tissue of disease part caused by stimulation and forms tumor, such as grape root cancer, cabbage root swelling, sakura crown gall, corn tumor smut, pepper root knot nematode, soybean cell cyst nematode, etc.

④Leaf curl: It refers to the curl and wrinkle of leaves. Sometimes the diseased leaves are thickened, hard, and become rolling when serious, such as peach leaf curl, potato leaf curl, broad bean leaf yellowing and curl, etc.

⑤Fern leaf: It refers to that leaves become filamentous, threadlike, or fern, such as tomato fern leaf, pepper fern leaf, etc.

⑥Deformity fruit: It refers to the deformation of fruit.

⑦Small fruit: It refers to that the diseased fruit is smaller than normal fruit.

⑧Phyllody: It refers to that floral organs of diseased plant become leaves, so that the plants can not bloom or bear fruit normally, which is also known as atavism of flowering organs, such as Chinese rose phyllody, peony phyllody, etc.

⑨Fasciation: It refers to that the branches of infected plants become flat, such as apple branch fasciation, cucumber stem fasciation, tamarisk branch fasciation, etc.

2. Sign Types of Plant Diseases

Signs refer to the special structures that are produced by pathogenic fungi (sometimes bacteria) on the surface of diseased plant tissues, such as vegetative bodies, reproductive bodies or disease products. Plant pathogenic bacteria only show the similar appearance of the signs, but there are obvious morphological differences between different types of pathogenic fungi, which

Root rot

Flower rot

Fruit rot

Dry rot

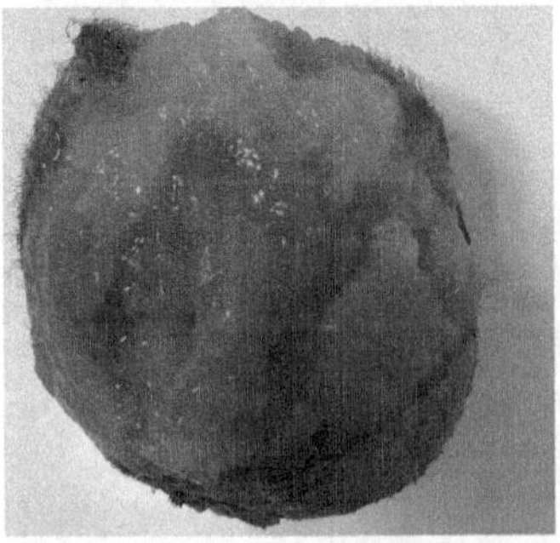
Soft rot

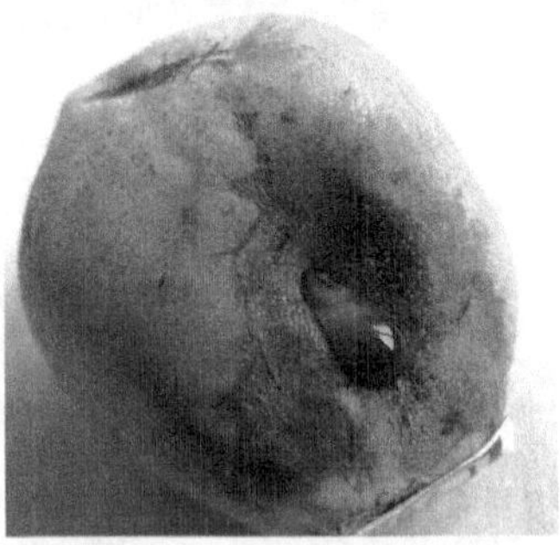
Wet rot

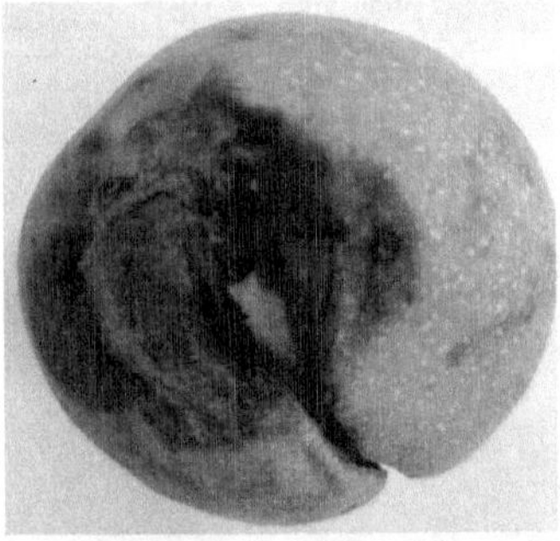
Black rot

Brown rot

White rot

Decay

Appended figure 1-4　Morbidities of Plant Diseases——Rot

Appended figure 1-5　Morbidities of Plant Diseases——Malformation

results in different signs. Common types of signs include powder, mould, particles, sclerotium, mushroom, bacteria ooze, etc.

1. Powder

Some plants can form a layer of powder on the surface of its infected leaf, branch or fruit, which can be divided into yellow rust, white powder, and black powder according to the shape

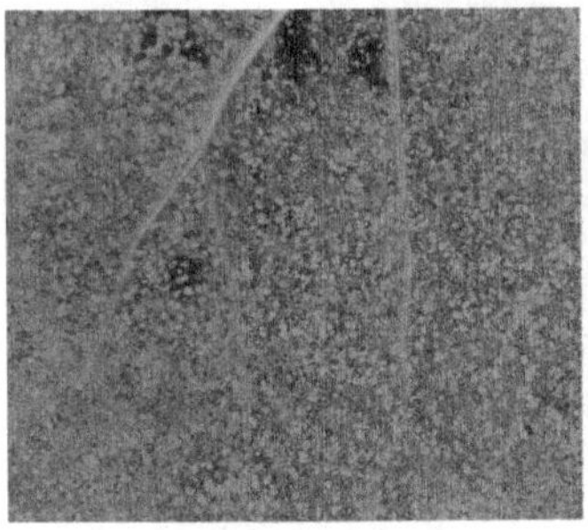

Yellow rust

White powder

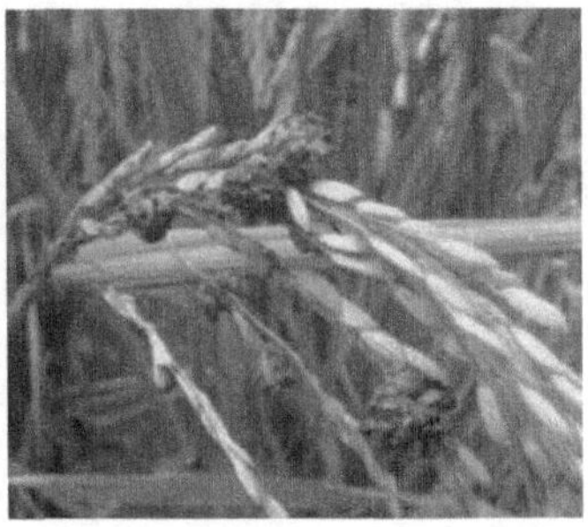

Black powder

Appended figure 1-6 Signs of Plant Diseases——Powder

and color of the powder layer(Appended figure 1-6).

①Yellow rust: The powder are different in colors, such as bright yellow, orange yellow, brown, or black brown, which is often the unique sign for plant rust.

②White powder: It refers to the white powdery substance produced on the surface of plant tissues, which is often the unique characteristic of powdery mildew. It is initially white, later turns brown, and produces numerous yellowish-brown globular particles (cleistothecia) that eventually turn black on the powder layer.

③Black powder: The powder is black in color, which usually occurs in the panicle, inside and outside of the grain, and inside the leaf or stem of the diseased plant. The number of black powder is very large, especially significant, which is the characteristic of the spore formation of the black powder fungus and the characteristic of the smut of various plants.

2. Mould

It is composed of mycelia, conidiospores, conidiophores, sporangia and sporangiophores of various fungi. The differences of color, shape, structure, and density of mold layer can mark different pathogens(Appended figure 1-7).

①Downy mildew: It is the special sign of plant downy mildew. In structure, the mould layer is composed of sporangiophores and sporangia of pathogens, which is mostly white in color, sometimes gray, purple or black.

②Hairy mould: It is a sign of some rotting diseases, with a rich mildew layer, which is white in the early stage, and then becomes black and white or with a dense layer of black globules (sporangia) on the surface in the late stage. It is mostly a sign of rot of fruits and tubers caused by *Mucor* or *Rhizopus*.

③Green mould: It occurs in a variety of diseased fruits, root crops, and tubers, which is green and mostly the sign of *Penicillium* infection.

④Gray mould: It can harm the fruits and floral organs of many plants. It is mainly caused by the genus *Botrytis*. The mould layer is composed of hyphae and conidia(conidiophore) of the pathogen.

Downy mildew

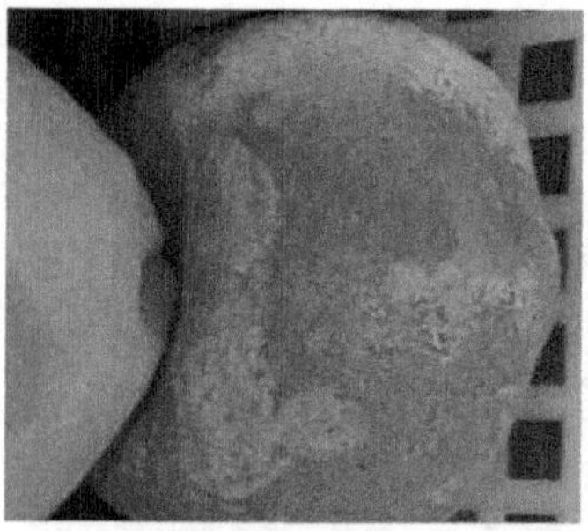

Green mould

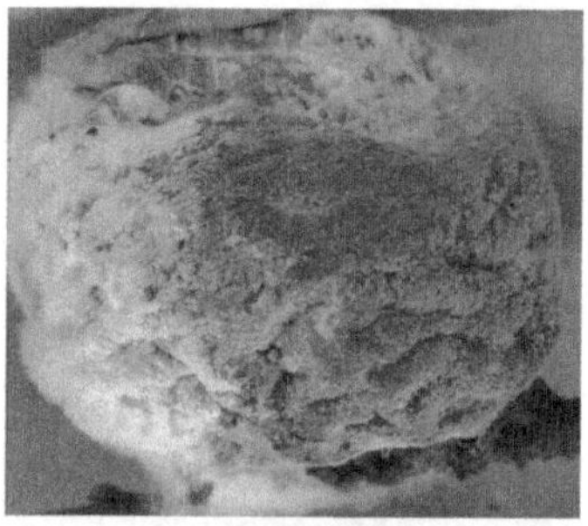

Gray mould

Hairy mould

Black mould

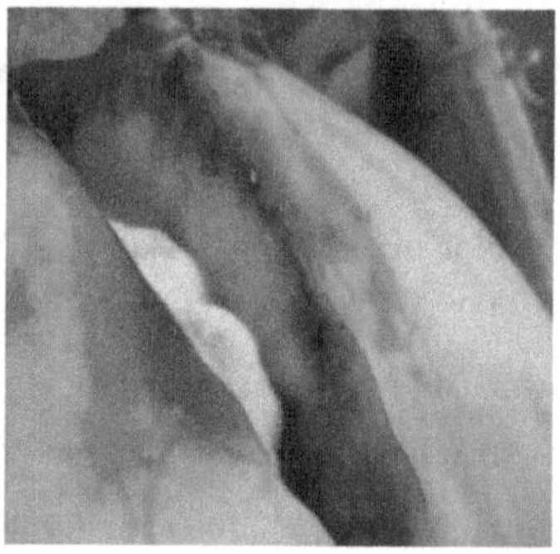

White mould

Appended figure 1-7 Signs of Plant Diseases——Mould

⑤Black mould: It is accompanied by a variety of diseases, mainly necrotizing diseases. The characteristic of mold layer is significantly different, which is loose or dense, generally thin. It is mainly a sign of the plant diseases caused by Anamorphic Fungi.

3. Particles

It refers to some small particles that can be produced on the surface of diseased tissues, especially stems and leaves, most of which are the fruiting bodies of the pathogen, such as stroma, cleistothecium, perithecium, pycnidium, acervulus, etc. The particles produced by different diseases are different in shape, size, color, protrusion degree, density and quantity. The common color of particles is black, blue, purple, orange, etc(Appended figure 1-8).

4. Sclerotium

The scelrotium is a dormant structure formed by the interleavings of fungal hyphae, most of which is black and a few is brown. It is often accompanied by whole or partial rot or necrosis, occurring on the surface of plant diseased tissues or in the pulp cavity of the stem. This type of plant disease is often called as sclerotiniose(Appended figure 1-9).

5. Mushroom

It usually refers to the fruiting bodies of higher Basidiomycota with various morphology on tree branches or decaying wood, such as pileus, lawn, and wood hoof-like structures, which can cause xylem dacay on the diseased trunks(Appended figure 1-9).

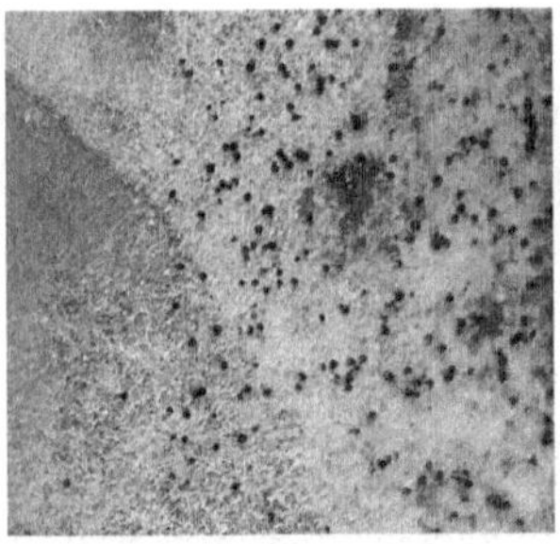
Cleistothecium

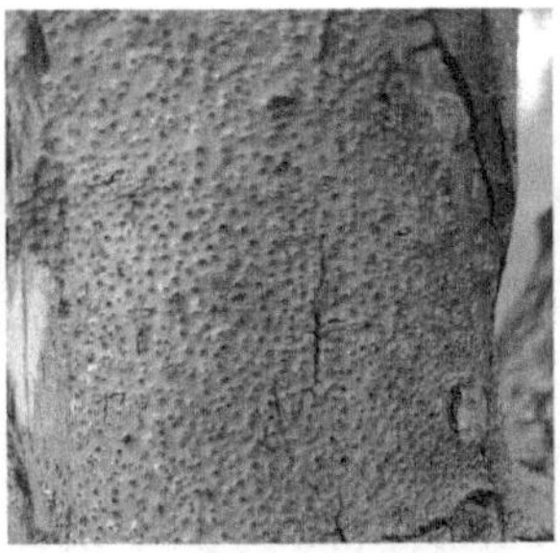
Perithecium

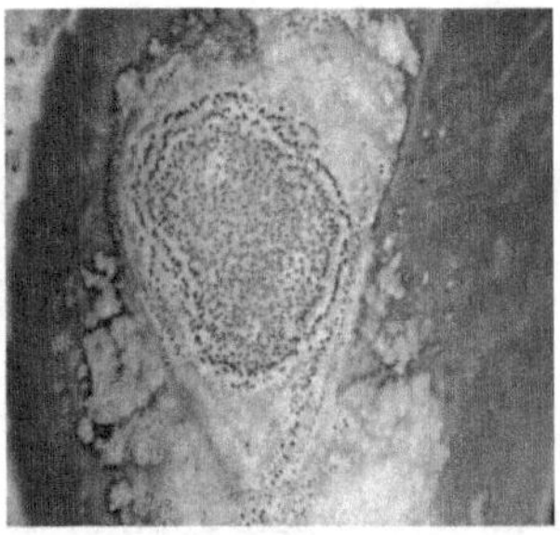
Acervulus

Stroma

Appended figure 1-8　Signs of Plant Diseases——Particles

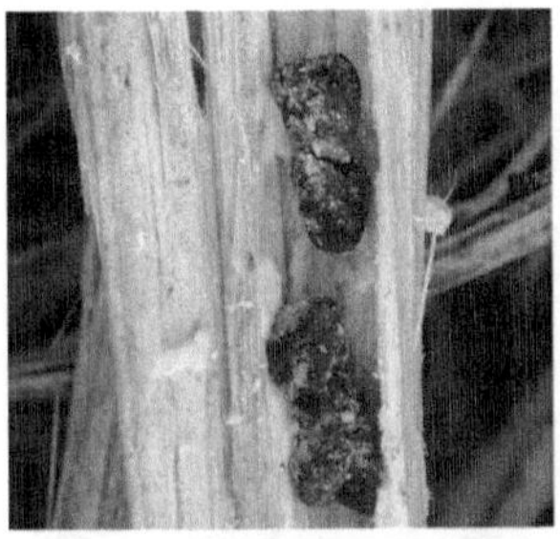
Sclerotium

Mushroom

Ooze(Cited from Zhigang Xu,2009)

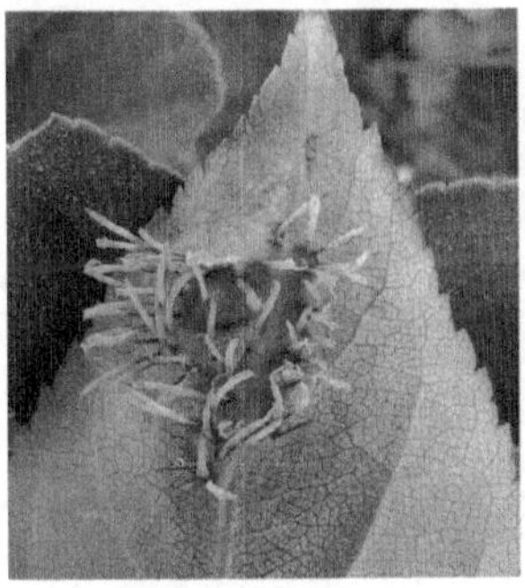
Aecium

Mycocecidium

Gummosis

Appended figure 1-9　Other Signs of Plant Diseases

6. Bacterial ooze

It is the special sign for most bacterial plant diseases, which contains a large number of bacterial individuals. The ooze is usually white, green, or yellow in colour, which will become bacterial colloidal particles or biofilms when dry(Appended figure 1-9).

7. Others

Rust fungi can form different shapes of aecia, such as hair columnar, tongue-like or cup-like, or form teliospore galls in the affected plant parts. Some fungal diseases can also form root-like rhizomorph or produce gummosis(Appended figure 1-9).

附录Ⅱ 临时显微玻片的制作

大部分植物病原物个体十分微小，需要借助显微镜才能观察到其具体的形态特征。在植物病理学实验中，要经常制作临时显微玻片来观察病原物在植物上着生的情况以及病原物的形态特征。临时显微玻片的制作都要使用浮载剂，并根据观察目标选用不同的制作方法。

1. 浮载剂

浮载剂是显微镜观察中必备的液体，其作用是防止观察材料的干燥变形和光线的散失，以利于显微镜观察。常用的浮载剂有以下几种：

①蒸馏水：是最常用的浮载剂，应用最为方便。对细菌、真菌孢子等无不利影响。观察细菌的喷菌现象、线虫活动以及真菌孢子萌发等都必须用水为浮载剂。测量真菌菌丝直径和真菌孢子大小时，以水为浮载剂最好。但是用水为浮载剂制片时较易形成气泡，且易蒸发干燥而不易保存玻片。为避免蒸发过快，可用稀释1倍的甘油代替。

②乳酚油：是应用最广泛的浮载剂，兼有透明防腐作用，可以较长期保存玻片。但乳酚油有使孢子膨大的作用，不能用作测量孢子大小。其配方为：乳酸20 mL，石炭酸(苯酚)20 mL，甘油40 mL，蒸馏水20 mL。如需染色，再向其中加入0.05~0.10 g的苯胺蓝或0.1 g酸性复红，即成乳酚油染剂，可使原生质着色。

③甘油明胶：其主要作用是黏着玻片以便长久保存，它既是黏着剂，又是浮载剂。配制时取白明胶5 g在30 mL水中加热融化，然后加入甘油35 mL、苯酚1 g搅拌均匀，纱布过滤后在玻瓶中保存，过滤时可在70~80℃的水浴锅中进行。过滤后在此温度下静止6~12 h以除去气泡。

使用时，先把封存的材料放在玻片中央，挑取甘油明胶一小团放在材料上，以盖玻片斜倚在胶团上，在酒精灯上轻微加热溶化，盖玻片即自动覆盖，或是把胶瓶在热水中加热融化后，再用玻璃棒加在载玻片上。

2. 临时玻片的制作

根据观察材料和观察目标的不同，可采用不同的方法制作临时玻片。常用的方法有以下几种：

①挑取法：采用挑取法可以直接用挑针从病组织或培养基上挑取表面的霉状物、粉状物或孢子团制成玻片。操作时要注意不能挑得太多，以免病原物繁殖体或营养体互相重叠导致观察不清。

②刮取法：对于病原体稀少、或用放大镜也不易清楚辨认霉层存在的病害标本，可采用刀片(蘸取少许浮载剂)刮取病原物繁殖体，要在病斑上顺同一个方向刮取2~3次，将刮得的霉状物转移至浮载剂中。玻片应擦净，浮载剂尽可能少，否则，这些微小的病原体在浮载剂中极易分散漂流，显微镜下难以寻找。

③撕取法：用小金属镊子撕下病部表皮或表皮毛制成临时玻片。该方法可以观察着生在寄主表面的菌丝和孢子，寄主表皮细胞内的真菌菌丝、吸器和休眠孢子囊堆，以及病毒的内含体。

④拔取法：把带有病原体的植物组织拨下，放入浮载剂中，用尖嘴镊子稳定材料，再用手术刀切去不含病原体的植物组织，使病原菌子实体外露，然后再在显微镜下观察。这类病原菌的繁殖器官多产生在植物表皮层下或体内，如子囊壳、子囊座、分生孢子器和分生孢子盘等。该方法制片只能看到病菌繁殖体的轮廓及孢子形态。

⑤粘贴法：将透明胶带剪成5 mm左右的正方形小块(注意胶带上不能带指印)，将胶面贴在病部，轻按后揭下制成玻片。该法适用于菌丝或子实体着生于病组织或培养基表面的材料，特别适用于观察分生孢子在分生孢子梗上的着生情况。

⑥徒手切片法：该方法是临时玻片制作最常用的方法之一。制成的玻片可以保持寄主组织和病原物原有的色泽，还可以观察病组织和病原物的解剖结构。切得好的徒手切片并不比石蜡切片差，而且非常方便。用树脂、指甲油或油漆封固后还可以作为半永久玻片保存。

选取病状典型、病征明显的病组织材料，在病征明显部位切取病组织小块(长方形较好，边长5~8 mm)，若为较硬而厚的材料，可将其放在载玻片上，用手指轻轻按住，随着手指的后退，用刀片将病组织小块切成若干很薄的细丝或片，用蘸有浮载剂的挑针挑选薄而合适的材料放在另一干净的载玻片上，制成临时玻片；若为薄而软的材料，可将病组织小块夹在胡萝卜、马铃薯或通草等固定材料中，连同固定物一起切成薄片，将切取物转移至盛有清水的小培养皿中，用尖头镊子轻轻搅动，使固定材料和病组织薄片分开，最后用尖头镊子选取薄而合适的切片，制成临时玻片。通草是最常用的固定材料，使用前应在70%乙醇中浸泡。

病组织材料若过于干燥，为防止切片时破碎，可先蘸取少量水湿润软化后再切。徒手切片是重要的基本操作，需反复练习。

⑦涂抹法：细菌和酵母菌的培养物常用该方法。将细菌或酵母菌的悬浮液均匀地涂抹在洁净的载玻片上，在酒精灯火焰上烘干、固定、再加盖玻片封固。加盖玻片之前还可以进行染色处理，使菌体或鞭毛着色而易于观察。

⑧组织透明法：将少量病组织材料切成细丝后放在载玻片上，滴加乳酚油后在酒精灯上缓缓加热至蒸汽出现。如此处理数次使植物组织透明，冷却后加盖玻片进行镜检，该方法可观察到病原物在寄主组织内的原有状态。

Appendix Ⅱ Preparation of Temporary Microscopic Slide

Most plant pathogenic individuals are extremely small that we need to observe their specific morphological characteristics using a microscope. In plant pathology experiments, temporary microslides are often made to observe the growth of pathogens on plants and the morphological characteristics of pathogens. Floating agents are used in the production of temporary microslides, and different methods can be used according to the observation targets.

1. Floating Agents

Floating agent is a necessary liquid for microscope observation. Its role is to prevent the deformation of the observation material due to dehydration and the loss of light during microscope observation. The commonly used floating agents are as follows.

①Distilled water: It is the most commonly used and most convenient floating agent. It has no adverse effect on bacteria, fungal spores, etc. It is necessary to use water as a floating agent to observe the bacterial spray, nematode activity, and fungal spore germination. However, when water is used as a floating agent, it is easy to form bubbles, and also easy to evaporate and dry. So it is not suitable for slide preservation. In order to avoid rapid evaporation, the water can be replaced with 50% glycerol.

②Lacteal phenol oil: Lacteal phenol oil is the most widely used floating agent. It is transparent and has antiseptic effect and can preserve glass slides for a long time. But lacteal phenol oil can enlarge spores and cannot be used to measure the size of spores. The formula is: 20 mL lactic acid, 20 mL carbolic acid (phenol), 40 mL glycerin, 20 mL distilled water. If staining is required, add 0.05-0.10 g aniline blue or 0.1 g acid complex red, and it becomes lacteal phenol oil dye.

③Glycerin gelatin: Its main role is to adhere to the glass slide for long-term preservation. It is both an adhesive and a floating agent. When preparing, heat and melt 5 g of gelatin in 30 mL water, and then add 35 mL glycerol and 1 g phenol and stir the solution evenly. After filtration, store the solution in a glass bottle. Filter the solution in a water bath at 70-80℃ and is incubate the filtrate at this temperature for 6-12 h to remove bubbles.

When use, first put the sealed material in the center of the glass slide, pick out a small amount of glycerin gelatin and put it on the material. Recline the coverslip on the gelatin and slightly heat it to melt on an alcohol lamp. The coverslip is automatically covered. Or heat the gel in plastic bottle to melt in hot water, and then add some gel to the glass slide.

2. Preparation of Temporary Slides

According to the different observation materials and observation targets, different methods can be used to make temporary glass slides. The common methods are as follows.

①Picking: The method of picking can be directly used to pick the mildew, powder or spore balls from the disease tissue or culture medium to make glass slides. During the operation, be careful not to pick too much, so as to avoid the overlap of pathogen propagules or vegetative bodies that make the observation not clear.

②Scraping: For the disease specimens with few pathogens or with mildew layer that is not easy to clearly identify with a magnifying glass, the fruiting bodies of them can be scraped with a blide (dipped in a little floating agent). Scrape two or three times in the same direction on the lesion, and transfer the scraped moldy material to the floating agent. The slides should be wiped clean and the floating agent as little as possible, otherwise, these tiny pathogens in the floating agent are very easy to disperse and drift, which might make it very difficult to observe under the microscope.

③Tearing: Tear off the epidermis or skin of the infected site with small metal tweezers and make a temporary slide. The hyphae and spores on the surface of the host plant, fungal hyphae, haustoria, and dormant spores in the epidermal cells of the host, and the inclusion bodies of the virus can be observed.

④Pulling: Pull the plant tissue with pathogen, put in the floating agent, stable the material with a pointed mouth tweezers, cut the tissue without pathogens using a knife, expose the pathogen fruit body, and then observe it under the microscope. The reproductive organs of these pathogens are mostly produced under the plant epidermis or in the body, such as perithecium, ascostroma, pycnidium, and acervulus, etc. Only the outline of the propagules and the spore morphology could be seen by this method.

⑤Pasting: Cut the scotch tape into square pieces of about 5 mm (no fingermarks on the adhesive tape), stick the adhesive surface on the infected site, gently press it and peel it off to make glass slides. This method is suitable for the material with hyphae or fruiting bodies growing on the surface of the diseased tissue or culture medium, especially for the observation of conidia growing on the conidiophores.

⑥Hand-making section: This method is one of the most commonly used methods for making temporary slides. The finished glass slides can maintain the original color of the host tissue and pathogens, and can also be used to observe the anatomical structure of the diseased tissue and pathogens. The hand-cutting section is no worse than the paraffin section, and it is very convenient to use. It can also be preserved as a semi-permanent glass slide after being sealed with resin, nail polish, or paint.

Select diseased tissue material with typical symptom and obvious signs, and cut small pieces (rectangle is better, 5-8 mm length) in the site with obvious sign. If the material is hard and thick, put it on the slide, hold it with your fingers gently, as the back of the finger, cut the diseased tissue pieces into very thin filaments or slices with the blade, select a thin, suitable material with a pick needle dipped in floating agent, place it on another clean glass slide, and make a temporary slide. For thin and soft material, clamp small piece of diseased tissue in fixed materials such as carrot, potato, or ricepaperplant pith, slice it along with the fixture, transfer the slices to a Petri dish containing water, gently stir them with a pointed tweezer to separate the fixture and the diseases tissue slices, and select thin and proper slices with the pointed tweezer to make temporary slides finally. Ricepaperplant pith is the most commonly used fixing material. It should be soaked in 70% alcohol before using.

If the diseased tissue material is too dry, make it moisten and soften by dipping it in a small amount of water before cutting to prevent broken. Freehand slicing is an important basic operation, which needs repeated practice.

⑦Smearing: This method is commonly used for bacteria and yeast culture. Smear bacteria or yeast suspension evenly on a clean glass slide, dry the suspention on alcohol lamp flame for fixing, and then cover a glass slide. Staining can be applied before covering, so that the bacteria or flagella can be colored and easy to observe.

⑧Tissue transparency method: Cut a small amount of diseased tissue material into filaments and place them on a glass slide, add a drop of lacteal phenol oil, heat the material slowly on an alcohol lamp until steam appears. Repeat this treatment for several times until the plant tissue becomes transparent, cover a glass slide after cooling and observe under a microscope. Using this method can observe the original state of the pathogen in host tissue.

附录 Ⅲ 植物病理学实验的常用培养基

1. 植物病原菌物常用培养基

(1)马铃薯葡萄糖琼脂培养基(PDA)

配方：马铃薯(去皮)200 g，葡萄糖 10~20 g，琼脂 17~20 g，蒸馏水 1 000 mL。

PDA 植物病理学中最常用的培养基，主要用于菌物的分离和培养，偶用于细菌的培养。

(2)马铃薯蔗糖琼脂培养基(PSA)

配方：马铃薯(去皮)200 g，蔗糖 10~20 g，琼脂 17~20 g，蒸馏水 1 000 mL。

(3)水琼脂培养基(WA)

配方：琼脂 10~17 g，蒸馏水 1 000 mL。

该培养基主要用于真菌的单孢分离及孢子萌发。

(4)燕麦培养基

配方：燕麦片 30 g，琼脂 20~30 g，水 1 000 mL。称取燕麦片后，加水 1 000 mL，煮沸 1 h 后，双层纱布过滤，收集滤液，补足水至 1 000 mL，加入琼脂。

该培养基可以促使某些菌物形成孢子和子实体，常用于多种疫霉菌的分离、培养和诱导产生卵孢子；稻瘟菌在其上也容易大量产孢。

(5)胡萝卜培养基

配方：新鲜胡萝卜 200 g 切成小片，加蒸馏水 500 mL，用组织捣碎机捣碎约 40 s，用 4 层纱布过滤去渣，补充水至 1 000 mL，加入琼脂 17~20 g。

该培养基较常用于多种疫霉菌的分离、培养和诱导产生卵孢子。

(6)马铃薯胡萝卜琼脂培养基(PCA)

配方：马铃薯(去皮)20 g，胡萝卜(去皮)20 g，琼脂 17~20 g，蒸馏水 1 000 mL。马铃薯和胡萝卜在 500 mL 水中煮沸 30 min，双层纱布过滤，加琼脂，补足水。

(7)玉米粉培养基

配方：玉米粉 30 g，水 1 000 mL，琼脂 18~20 g。配制方法与燕麦培养基相似。

(8)番茄培养基

配方：番茄汁 20 mL，$CaCO_3$ 0.4 g，琼脂 18~20 g，蒸馏水 80 mL。

番茄汁的制备方法：取新鲜成熟番茄果实，用自来水洗净切成片，置于组织捣碎机中匀浆 2 min，经双层纱布过滤种子和组织残余，过滤液即为所制备的番茄汁。

该培养基较常用于多种疫霉菌的培养与产孢，也较适用于其他真菌产孢。

(9)大豆培养基

配方：大豆汁 10 mL，琼脂 2 g，蒸馏水 90 mL。

大豆汁的制备方法：60 g 干大豆种子用水冲洗净，浸泡过夜，与 330 mL 蒸馏水混合，用组织捣碎机破碎 2 min，经单层纱布过滤去渣，滤液即为所制备的大豆汁。

该培养基较常用于多种疫霉菌的分离、培养与产孢。

(10)黑麦培养基

配方：黑麦 50 g，琼脂 20 g，蒸馏水 1 000 mL。取 50 g 黑麦种子在 1 000 mL 蒸馏水中浸泡 24~36 h，用组织捣碎机破碎 2 min，经 4 层纱布过滤去渣，滤过液补足水至 1 000 mL，加入琼脂 20 g，在 121℃下高压蒸汽灭菌 30 min。

(11)利马豆培养基

配方：利马豆粉 60 g，琼脂 20 g，水 1 000 mL。

利马豆粉 60 g，加水 1 000 mL，在 60℃下水浴 1 h，双层纱布过滤去渣，收集滤液并补足水至 1 000 mL，加入琼脂 20 g，加热煮沸 25~30 min。

该培养基适用于大多数疫霉菌的分离、培养、繁殖和保存。但不适用于培养马铃薯晚疫病菌。该培养基在加热过程中会产生大量气体，因此在灭菌前应充分煮沸。

(12)V8 培养基

该培养基是用美国 Campbell 公司生产的以 8 种蔬菜为主要成分混合制成的罐装 V8 汁作为主要原料配制而成的。对于某些生长慢且不易产孢的尾孢属真菌等，用此培养基可使其生长速率快，易产孢。用于疫霉时，因不同的实验要求，需要采用不同的配比，可配置成以下几种培养基。

①10% V8 培养基：V8 汁 10 mL，蒸馏水 90 mL，$CaCO_3$ 0. 02 g，琼脂 2 g。该培养基常用于疫霉菌分离、培养、保存、诱导孢子囊产生、交配型测定和产生卵孢子。

②5% V8 培养基：V8 汁 5 mL，蒸馏水 90 mL，$CaCO_3$ 0. 02 g，琼脂 2 g。该培养基常用于疫霉菌分离、培养和保存。

③10% V8 培养液：V8 汁 100 mL，$CaCO_3$ 1 g，主要用于诱导疫霉产生孢子囊。

(13)查彼(Czapek)培养基

配方：$NaNO_3$ 2 g，K_2HPO_4 1 g，KCl 0. 5 g，$MgSO_4 \cdot 7H_2O$ 0. 5 g，$FeSO_4$ 0. 01 g，蔗糖 30 g，琼脂 15~18 g，蒸馏水 1 000 mL。

(14)理查(Richard)培养基

配方：KNO_3 10 g，KH_2PO_4 2. 5 g，$MgSO_4 \cdot 7H_2O$ 2. 5 g，$FeCl_3$ 0. 02 g，蔗糖 50 g，琼脂 15~18 g，蒸馏水 1 000 mL。

2. 植物病原细菌常用培养基

(1)营养琼脂培养基(NA)

配方：牛肉浸膏 3 g，蛋白胨 5~10 g，葡萄糖 5~10 g，酵母粉 1 g，琼脂 15~18 g，蒸馏水 1 000 mL。

该培养基是细菌最常用的培养基，又称为肉汤培养基，简称 NA 培养基。主要用于细菌的分离、培养、纯化和保存。

(2)牛肉膏蛋白胨培养基

配方：牛肉浸膏 5 g，蛋白胨 10 g，NaCl 5 g，琼脂 15~20 g，蒸馏水 1 000 mL，pH 7. 2。

最常用的细菌培养基，常用来代替 NA 培养基。主要用于细菌的培养，也可用于细菌的生理生化测定。

(3)Luria-Bertani 培养基(LB)

配方：胰蛋白胨 10 g，酵母粉 5 g，NaCl 10 g，琼脂 16~18 g，水 1 000 mL，pH

7. 2~7. 4。

该培养基营养比较充分，简称 LB 培养基。主要用于欧氏杆菌、假单胞杆菌、黄单胞杆菌等细菌的分离、培养和保存。

(4)酵母浸膏葡萄糖碳酸钙培养基(YDC)

配方：酵母浸膏 10 g，葡萄糖 20 g，碳酸钙 20 g，琼脂 15~18 g，蒸馏水 1 000 mL。

碳酸钙应充分研磨，培养基在装管、摆斜面或倒平板前应摇匀，使碳酸钙与其他成分充分混匀，避免其快速沉淀。

(5)金氏 B 培养基(KB)

配方：蛋白胨 20 g，甘油 10 mL，$K_2HPO_4 \cdot 3H_2O$ 2. 5 g，$MgSO_4 \cdot 3H_2O$ 1. 5 g，琼脂 15~18 g，蒸馏水 1 000 mL。

该培养基主要用于分离、鉴定荧光假单胞杆菌。

(6)D1 培养基

配方：甘露醇 15 g，$MgSO_4 \cdot 7H_2O$ 0. 2 g，$NaNO_3$ 5 g，LiCl 6 g，K_2HPO_4 2 g，溴麝香草酚蓝 0. 1 g，琼脂 15~18 g，蒸馏水 1 000 mL。121℃灭菌 15 min，灭菌后调培养基 pH 值至 7. 0。

该培养基主要用于分离、培养和区分土壤杆菌。

(7)柠檬酸钠培养基

配方：NaCl 5 g，$MgSO_4 \cdot 7H_2O$ 0. 2 g，$NH_4H_2PO_4$ 1 g，K_2HPO_4 1 g，柠檬酸钠 2 g，琼脂 20 g，溴麝香草酚蓝 1%(*W/V*)溶液(溶于 50%乙醇)15 mL，蒸馏水 1 000 mL。

将培养基的 pH 值调到 6. 8，分装试管后灭菌，斜面凝固后接菌，培养 24~48 h，如果柠檬酸钠被利用，则培养基变成蓝色。

(8)高糖培养基

配方：蔗糖 160 g，0. 1%的放线菌酮 20 mL，1%的结晶紫(乙醇溶液)0. 8 mL，琼脂 12 g，蒸馏水 380 mL。

该培养基主要用于梨火疫病病原菌的培养。

(9)TTC 培养基

配方：葡萄糖 10 g(或甘油 5 mL)，酪朊水解物 1 g，蛋白胨 10 g，琼脂 18 g，蒸馏水1 000 mL。

每瓶分装上述培养基 200 mL，常规灭菌。在倒培养皿前，每瓶加 1%的 2，3，5-三苯基氯化四氮唑(TTC)溶液 1 mL(1%的 TTC 溶液 115℃灭菌 7~8 min，置于低温暗处备用)。

(10)SX 培养基

配方：可溶性淀粉 10 g，甲基紫 2B 1 mL，牛肉浸膏 1 g，1%的甲基绿 2 mL，NH_4Cl 5 g，琼脂 15~18 g，蒸馏水 1 000 mL，pH 7. 0~7. 1。

该培养基主要用于植物病原细菌的培养。

3. 植物病原线虫常用培养基

(1)麦芽琼脂培养基

配方：麦芽粉 20 g，琼脂 20~30 g，蒸馏水 1 000 mL，pH 7. 2。

该培养基常用于小杆属线虫和一些植物寄生线虫的培养。

(2)卵磷脂琼脂培养基

配方：K_2HPO4 0.75 g，$MgSO_4 \cdot 7H_2O$ 0.75 g，KNO_3 3 g，NaCl 2.75 g，卵磷脂 1 g，酵母粉 1 g，蛋白胨 2.5 g，琼脂 20~30 g，蒸馏水 1 000 mL，pH 7.2。

该培养基可用来培养一些兼性寄生的小杆属线虫和垫刃属线虫。

(3)用真菌培养线虫

利用真菌是进行一些植物寄生线虫人工繁殖的主要方法，应用最多的真菌是灰葡萄孢。常用玉米粉琼脂培养基培养灰葡萄孢菌，待平板上长满菌丝后，将线虫接在真菌平板上培养。松材线虫就常采用灰葡萄孢进行人工培养。

4. 土壤微生物常用培养基

(1)高氏一号培养基

配方：可溶性淀粉 20 g，K_2HPO_4 0.5 g，$MgSO_4 \cdot 7H_2O$ 0.5 g，KNO_3 1 g，NaCl 0.5 g，$FeSO_4$ 0.01 g，琼脂 20 g，蒸馏水 1 000 mL，pH 7.2~7.4。

该培养基主要用于培养放线菌和一些土壤微生物。

(2)土壤浸渍液琼脂培养基

配方：土壤浸渍液 100 mL，琼脂 15~20 g，蒸馏水 1 000 mL，pH 7.0。称 1 kg 土壤，加 1 000 mL 水，高压蒸汽(121℃)处理 30 min，用双层滤纸过滤，收集滤液，再加入琼脂，加热至溶化，补足失水后，分装灭菌。

该培养基可用于许多土壤微生物的分离与培养。

Appendix Ⅲ　Common Media for Experiments of Plant Pathology

1. Media Commonly Used for Plant Pathogenic Fungi

(1) Potato Glucose Agar Medium (PDA)

Formula: 200 g potato (peeled), 10-20 g glucose, 17-20 g agar, 1 000 mL distilled water.

PDA is the most commonly used medium in plant pathology, mainly used for the isolation and culture of fungi, occasionally for the culture of bacteria.

(2) Potato Sucrose Agar Medium (PSA)

Formula: 200 g potato (peeled), 10-20 g sucrose, 17-20 g agar, 1 000 mL distilled water.

(3) Water Agar Medium (WA)

Formula: 10-17 g agar, 1 000 mL distilled water.

This medium is mainly used for single spore isolation and spore germination of fungi.

(4) Oat Medium

Formula: 30 g oatmeal, 20-30 g agar, 1 000 mL distilled water. Weigh the oatmeal, add 1 000 mL water and boil it for 1 h, filter with double-layer of cheese cloth, collect the filtrate, add water to 1 000 mL, and then add agar.

The medium can promote the formation of spores and fruiting bodies of some fungi, and is often used for isolation, culture, and oospore induction of a variety of *Phytophthora*. It is also easy for rice blast fungus to produce a large number of spores.

(5) Carrot Medium

Formula: Cut 200 g fresh carrot into small pieces, add 500 mL distilled water, mash it with tissue masher for about 40 s, filter it with 4 layers of cheese cloth to remove residue, add water to 1 000 mL, and then add 17-20 g agar.

This medium is commonly used for isolation, culture, and oospore induction of a variety of *Phytophthora*.

(6) Potato Carrot Agar Mmedium (PCA)

Formula: 20 g potato (peeled), 20 g carrot (peeled), 17-20 g agar, 1 000 mL distilled water. Boil potato and carrot in about 500 mL water for 30 min, filter with double-layer cheese cloth, add agar, and add water to 1 000 mL.

(7) Corn Meal Medium

Formula: 30 g corn meal, 1 000 mL water, 18-20 g agar. The preparation method is similar to oat medium.

(8) Tomato Culture Medium

Formula: 20 mL tomato juice, 0.4 g $CaCO_3$, 18-20 g agar, 80 mL distilled water.

Preparation of tomato juice: Take fresh ripe tomato fruit, wash with tap water and cut into pieces, homogenate for 2 min, filter through double-layer cheese cloth to remove seeds and tissue residue.

This medium is commonly used for the culture and sporulation of a variety of *Phytophthora*, but also suitable for the sporulation of other fungi.

(9) Soybean Medium

Formula: 10 mL soybean juice, 2 g agar, 90 mL distilled water.

Preparation of soybean juice: Wash 60 g dry soybean soak it in water overnight, mix it with 330 mL distilled water, crush it by a tissue masher for 2 min, filter the mixture through a single layer cheese cloth to remove residue. The filtrate is the soybean juice.

This medium is commonly used for isolation, culture, and sporulation of a variety of *Phytophthora*.

(10) Rye Medium

Formula: 50 g rye, 20 g agar, 1 000 mL distilled water.

Soak 50 g rye in 1 000 mL distilled water for 24-36 h and crush it for 2 min in a tissue masher. Remove the residue by filteration through 4 layers of cheese cloth, add water to the filtrate to 1 000 mL and add 20 g of agar. Autoclaved at to 121℃ for 30 min.

(11) Lima Bean Medium

Formula: 60 g Lima bean powder, 20 g agar, 1 000 mL water.

Add 60 g of lima bean powder to 1 000 mL of water, incubate it in a water bath at 60℃ for 1 h. After that, remove the residue by filtration through a double-layer cheese cloth, collect the filtrate and add water to 1 000 mL, and then add 20 g agar, which is then heated to boiling for 25-30 min.

This medium is suitable for the isolation, cultivation, reproduction, and preservation of most fungi in *Phytophthora* but not for the cultivation of *P. infestans*. The medium produces a large amount of gas during heating and should therefore be fully boiled before autoclaved.

(12) V8 Medium

This medium is prepared by canned V8 juice containing 8 kinds of vegetables produced by Campbell Company in the United States as the main ingredient. For some *Cercospora* fungi that grow slowly and sporulate difficultly, this medium can promote their growth and sporulation. When used in *Phytophthora*, the following media can be prepared due to different experimental requirements and different proportions.

①10% V8 medium: 10 mL V8 juice, 90 mL distilled water, 0.02 g $CaCO_3$, 2 g agar. This medium is commonly used for isolation, culture, preservation, sporangia induction, mating type determination and oospores production of *Phytophthora*.

②5% V8 medium: 5 mL V8 juice, 90 mL distilled water, 0.02 g $CaCO_3$, 2 g agar. This medium is often used for isolation, culture and preservation of *Phytophthora*.

③10% V8 medium: 100 mL V8 juice, 1 g $CaCO_3$. The medium is mainly used to induce sporangium of *Phytophthora*.

(13) Czapek medium

Formula: 2 g $NaNO_3$, 1 g K_2HPO_4, 0.5 g KCl, 0.5 g $MgSO_4 \cdot 7H_2O$, 0.01 g $FeSO_4$, 30 g sucrose, 15-18 g agar, 1 000 mL distilled water.

(14) Richard medium

Formula: 10 g KNO_3, 25 g KH_2PO_4, 2.5 g $MgSO_4 \cdot 7H_2O$, 0.02 g $FeCl_3$, 50 g sucrose, 15-18 g agar, 1 000 mL distilled water.

2. Media Commonly Used for Plant Pathogenic Bacteria

(1) Nutritional Agar Medium (NA)

Formula: 3 g beef extract, 5-10 g peptone, 5-10 g glucose, 1 g yeast powder, 15-18 g agar, 1 000 mL distilled water.

It is the most commonly used medium for bacteria, also known as broth medium, or NA medium, mainly used for the isolation, culture, purification, and preservation of bacteria.

(2) Beef Extract Peptone Medium

Formula: 5 g beef extract, 10 g peptone, 5 g NaCl, 15-20 g agar, 1 000 mL distilled water, pH 7.2.

It is the most commonly used bacterial medium, commonly used in place of NA medium. It is mainly used for the culture of bacteria, and for the physiological and biochemical determination of bacteria as well.

(3) Luria-Bertani Medium (LB)

Formula: 10 g peptone, 5 g yeast powder, 10 g NaCl, 16-18 g agar, 1 000 mL water, pH

7. 2-7. 4.

This medium has sufficient nutrition. The abbreviation is LB medium. It is mainly used for the isolation, culture, and preservation of *Enterobacter*, *Pseudomonas*, *Xanthomonas*, ect.

(4) Yeast Extract Dextrose Calcium Carbonate Medium (YDC)

Formula: 10 g yeast extract, 20 g dextrose, 20 g calcium carbonate, 15-18 g agar, 1 000 mL distilled water.

Calcium carbonate should be fully ground, and the medium should be shaken well before the loading into tubes or plates, so that the calcium carbonate can be fully mixed with other ingredients to avoid precipitation.

(5) King's B Medium (KB)

Formula: 20 g peptone, 10 mL glycerin, 2. 5 g $K_2HPO_4 \cdot 3H_2O$, 1. 5 g $MgSO_4 \cdot 3H_2O$, 15-18 g agar, 1 000 mL distilled water.

This medium is mainly used to isolate and identify *Pscudomonas fluorescens*.

(6) D1 Medium

Formula: 15 g mannitol, 0. 2 g $MgSO_4 \cdot 7H_2O$, 5 g $NaNO_3$, 6 g LiCl, 2 g K_2HPO_4, 0. 1 g bromothymol blue, 15-18 g agar, 1 000 mL distilled water. Adjust the pH to 7. 0 after autoclaving at 121℃ for 15 min.

It is This medium is mainly used for the isolation, culture, and distinguishing *Agrobacterium*.

(7) Sodium Citrate Medium

Formula: 5 g NaCl, 0. 2 g $MgSO_4 \cdot 7H_2O$, 1 g $NH_4H_2PO_4$, 1 g K_2HPO_4, 2 g sodium citrate, 20 g agar, 15 mL of 1% (*W/V*) bromothymol blue solution (dissolved in 50% alcohol), 1 000 mL distilled water.

The medium pH was adjusted to 6. 8. Autoclaved after dispensing the medium into test tubes, inoculate after the inclined surface solidified, and culture for 24-48 h. If sodium citrate is used, the medium turns blue.

(8) High Sugar Medium

Formula: 160 g sucrose, 20 mL of 0. 1% cycloheximide, 0. 8 mL of 1% crystal violet (alcohol solution), 12 g agar, 380 mL distilled water.

The medium is mainly used for the culture of *Erwinia amylovory*.

(9) TTC Medium

Formula: 10 g Glucose (or 5 mL glycerin), 1 g casein hydrolysate, 10 g peptone, 18 g agar, 1 000 mL distilled water.

Dispense 200 mL of the above medium into each bottle, autoclaved. Before pouring the petri dishes, add 1 mL of 1% 2, 3, 5-triphenyl tetrazole chloride (TTC) solution to each bottle (1% TTC solution is autoclaved at 115℃ for 7-8 min and store at in a cool and dark place).

(10) SX Medium

Formula: 10 g soluble starch, 1 mL methyl purple 2B, 1 g beef extrusion, 2 mL of 1%

methyl green, 5 g NH_4Cl, 15-18 g agar, 1 000 mL distilled water, pH 7. 0-7 1.

This medium is mainly used for the culture of plant pathogenic bacteria.

3. Media Commonly Used for Plant Pathogenic Nematodes

(1) Malt Agar Medium

Formula: 20 g malt powder, 20-30 g agar, 1 000 mL distilled water, pH 7. 2.

This medium is commonly used for the culture of nematodes of the genus *Rhabditis* and some plant parasitic nematodes.

(2) Lecithin Agar Medium

Formula: 0. 75 g K_2HPO_4, 0. 75 g $MgSO_4 \cdot 7H_2O$, 3 g KNO_3, 2. 75 g NaCl, 1 g lecithin, 1 g yeast powder, 2. 5 g peptone, 20-30 g agar, 1 000 mL distilled water, pH 7. 2.

The culture medium can be used for the culture of some facultative parasitic nematodes of *Rhabditis* and *Tylenchus*.

(3) Cultivation of Nematode Using Fungi

Using fungi to cultivate some plant parasitic nematodes is the main method for the artificial propagation of nematodes. The most commonly used fungus is *Botrytis cinerea*, which is usually cultured on corn meal agar medium. When the plate is covered with mycelia, the nematodes are inoculated on the plate for culture. *Bursaphelenchus xylophilus*, the pathogen of pine wilt nematode, is usually cultured using *B. cinerea*.

4. Medium Commonly Used for Soil Microorganisms

(1) Cauze's Medium

Formula: 20 g soluble starch, 0. 5 g K_2HPO_4, 0. 5 g $MgSO_4 \cdot 7H_2O$, 1 g KNO_3, 0. 5 g NaCl, 0. 01 g $FeSO_4$, 20 g agar, 1 000 mL distilled water L, pH 7. 2-7. 4.

This medium is mainly used for the culture of actinomycetes and some soil microorganisms.

(2) Soil Extract Agar Medium

Formula: 100 mL soil extract solution, 15-20 g agar, 1 000 mL distilled water, pH 7. 0.

Weigh 1 kg soil and add it into 1 000 mL water, which is then autoclaved at 121℃ for 30 min and filtered through a double-layer filter paper. The filtrate is collected and then add agar into it, which is then heated until dissolved, and add water to 1 000 mL for later subpackage and sterilization.

This medium can be used to isolate and cultivate many soil microorganisms.

附录Ⅳ 常用专业术语的中英文对照表
AppendixⅣ Glossary of Important Terms(English-Chinese)

A
acervulus 分生孢子盘
acquired resistance 获得抗病性
acquisition period 获毒期
active resistance 主动抗病性
aeciospore 锈孢子
aecium 锈孢子器
anamorph 无性态
anastomosis 菌丝融合
antheridium 雄器
anthracnose 炭疽病
Aphelenchoid oesophagi 滑刃型食道
apical paraphysis 顶侧丝
apothecium 子囊盘
appendage 附属丝
appressorium 附着胞
arthrospore 节孢子
artificial inoculation 人工接种
ascocarp 子囊果
ascogenous hypha 产囊丝
ascogonium 产囊体
ascospore 子囊孢子
ascostroma 子囊座
ascus 子囊
aseptate hypha 无隔菌丝
asexual reproduction 无性繁殖
autoclaving 高压蒸汽灭菌
autoecism 单主寄生
avoidance 避病性
B
bacteria exudation 喷菌现象
bacteria 细菌
bacteriophage 噬菌体
base pair 碱基对
basidiocarp 担子果
basidiospore 担孢子
basidium 担子
biological control 生物防治
biotroph 活体营养型
biovar 生化变种
bioassay 生物测定
biotype 生物型
bitunicate 双层壁的
blastic 芽殖
blight 疫病
blue-strain fungi 蓝变真菌
bunt 腥黑粉病
C
canker 溃疡
cause of disease 病因
chlamydospore 厚垣孢子
circulative period 循回期
circulative 循回型关系
clamp connection 锁状联合
cleistothecium 闭囊壳
colonization 定植
colony 菌落
color breaking 碎锦
columella 囊轴
commensalism 共栖
complex symptoms 并发症
conidiocarp 分生孢子果
conidiophore 分生孢子梗
conidium/conidiospore 分生孢子
cross protection 交叉保护
crown gall 冠瘿病
crozier 产囊丝钩
cultivation 培养
cultural control 农业防治
cystospore 休止孢
cyst 胞囊
D
damping off 猝倒
die-back 梢枯
differential host 鉴别寄主
diplanetism 双游现象
discoloration 变色
disease cycle 病害循环
disease index 病情指数
disease progress curve 季节流行曲线
disease severity 病害严重
disease tetrahedron 病害四面

disease triangle 病害三角
downy mildew 霜霉病
dwarf 矮化

E

economic threshold 经济阈值
effector 效应子
egg sac 卵囊
elicitor 激发子
ELISA 酶联免疫吸附反应
epiphyte 附生植物
Eukaryotes 真核生物
eu-form rust 全锈型
evocatory plant 诱发植物
exponential phase 指数增长期

F

facultative parasitism 兼性寄生
fasciation 带化
fern leaf 蕨叶
forecasting 预报
fruiting body 子实体
forma specialis（f. sp.）专化型
fungi 菌物/真菌
fungicide 杀真菌剂
Fungi Imperfecti 半知菌

G

gametangium 配子囊
gamete 配子
gene-for-gene theory 基因对基因学说
genotype 基因型
Gram stain 革兰氏染色
gray mold 灰霉病
gum 胶质
gummosis 流胶现象

H

haustorium 吸器
hemi-form rust 半锈型
hemiparasite 半寄生物
hemiparasitism 半寄生
heteroecism 转主寄生
heterokaryon 异核体
heterokaryosis 异核现象
heterothallism 异宗配合
holoparasitism 全寄生
homothallism 同宗配合
host 寄主
host specific toxin 寄主专化性毒素
hydathodes 水孔
hymenium 子实层
hyperplasia 增生
hypertrophy 增大
hypersensitivity 过敏反应
hypha 菌丝
hyphopodium 附着枝
hypoplasia 减生

I

incidence 发病率
inclusion body 内含体
incubation period 潜育期
indexing 检索
inducting resistance 诱发抗病性
infection process 侵染过程
infectious disease 侵染性病害
inhibition rate 抑制率
inoculation 接种
integrated control 综合防治
intercellular 细胞间的
intracellular 细胞内的
in vitro 离体
in vivo 活体
isolation 分离

K

karyogamy 核配
Koch's rule 科赫法则

L

latent infection 潜伏侵染
leaf scard 叶烧
leaf-curl 缩叶病
leaf spot 叶斑
lenticel 皮孔
life cycle 生活史
local lesion 局部坏死斑
locule 子囊腔
logistic phase 逻辑斯蒂增长期
long life-cycle 长生活史型

M

major gene resistance 主效基因抗病
malformation 畸形
masking of symptom 隐症现象
mechanical inoculation 机械接种
medium 培养基

meiosis 减数分裂
metamorphosis 变态
microscope slide 显微玻片
microscope 显微镜
microscopy 显微镜检
minor gene resistance 微效基因抗病性
monocyclic disease 单循环病害
monopartite virus 单分体病毒
monoplanetism 单游现象
morbidity 病状
mosaic 花叶
mottle 斑驳
multipartite virus 多分体病毒
multiplication 复制增殖
mycelium 菌丝体
mycotoxin 真菌毒素

N

necrosis 坏死
necrotic hypersensitive 过敏性坏死反应
necrotroph 死体营养型
nematode 线虫
nerve ring 神经环
networks loops 菌网
non-host resistance 非寄主抗病性
noninfectious disease 非侵染性病害
nutrition deficiencies 缺素症

O

obligate parasite 专性寄生物
obligate parasitism 专性寄生
oogonium 藏卵器
oosphere 卵球
oospore 卵孢子
ooze 菌脓

P

papilla 乳突
paraphysis 侧丝
paraphysoid 拟侧丝
parasexuality 准性生殖
parasite 寄生物
parasitic plants 寄生性植物
parasitism 寄生性
pathogenesis-related protein 病程相关蛋白
passive resistance 被动抗病性
pathogenesis 病程
pathogenicity 致病性
pathogen 病原物
pathovar 致病变种
penetration peg 侵入钉
penetration period 侵入期
penicilliosis 青霉病
perfect stage 有性阶段
perineal pattern 会阴花纹
periphysis 缘丝
periphysoid 拟缘丝
perithecium 子囊壳
pesticide 农药
pesudoperithecium 假子囊壳
phyllody 花变叶
physical control 物理防治
phytoalexin 植物保卫素
Phytoplasma 植原体
plant disease epidemics 植物病害流行
plant disease 植物病害
plant quarantine 植物检疫
plasmid 质粒
plasmodesmata 胞间连丝
plasmodium 原质团
plasmogamy 质配
polyetic 积年流行的
polycyclic disease 多循环病害
polymerase chain reaction 聚合酶链式反应
polymorphism 多型现象
prediction 预测
primary infection 初侵染
primary mycelium 初生菌丝体
probasidium 原担子
Prokaryotes 原核生物
proiferation 层出现象
promycelium 先菌丝
prosenchyma 疏丝组织
pseudofungi 假菌
pseudomycelium 假菌丝
pseudoparenchyma 拟薄壁组织
pustule 孢子堆
pycnidium 分生孢子器
pycniospore 性孢子
pycnium 性子器

Q

quarantine regulations 检疫法规

R

race-nonspecific resistance 非小种专化抗性
race-specific resistance 小种专化抗病性
race 小种
resistance 抗病性
resting sporangium 休眠孢子囊
rhizogen inducing plasmi 致发根质粒
rhizoid 假根
rhizomorph 菌索
ring spot 环斑
root knot 根结病
root tumer 根癌病
rot 腐烂
rust 锈病

S

saprogen 腐生物
satellite RNA 卫星 RNA
scab 黑星病
sclerotiniose 菌核病
sclerotium 菌核
secondary infection 再侵染
secondary mycelium 次生菌丝体
septate hypha 有隔菌丝
septum 隔膜
seta 刚毛
sexual compatible 性亲和
sexual incompatible 性不亲和
sexual reproduction 有性生殖
sexual spore 有性孢子
shot hole 穿孔
short-form rust 短锈型
short life-cycle 短生活史型
sign 病征
smut 黑粉病
soft rot 软腐
soil inhabitant 土壤习居菌
soil invader 土壤寄居菌
sooty mould 煤污病
sorus 孢子团
specialized parasitism 寄生专化性
species 种
spermatization 授精作用
Spiroplasma 螺原体
sporangiophore 孢囊梗
sporangiospore 孢囊孢子
sporangium 孢子囊
spore 孢子
sporodochium 分生孢子座
spot 叶斑
sterigma 小梗
stolon 匍匐菌丝
stroma 子座
stunt 矮缩
strain 菌株
stylet 口针
suberization 木栓化
subspecies 亚种
susceptible 感病
symbiosis 共生
symdrome 综合症
symptom 症状
symptom appearance 发病期
symptomless carrier 无症带毒者
systemic infection 系统侵染

T

taxonomy 分类学
teleomorph 有性态
teliospore 冬孢子
telium 冬孢子堆
tertiary mycelium 三生菌丝体
thallus 营养体
tinsel 尾鞭
tolerance 耐病性
toxin 毒素
tumor inducing plasmid 致瘤质粒
tylenchoid oesophagi 垫刃型食道
tylose 侵填

U

urediospore 夏孢子
uredium 夏孢子堆

V

vector transmission 介体传播
vector 介体
vegetative body 营养体
vein cleaning 脉明
vesicle 泡囊
vessel 导管
vetical resistance 垂直抗病性
virus 病毒
virion 病毒颗粒

viroids 类病毒
virulence 毒性
virulence gene 毒性基因
virulence factor 毒性因子
virulent 有毒的
viruliferous 带毒的
virusoid 拟病毒

W

whiplash 茸鞭
white powdery mildew 白粉病
wilt 萎蔫
withes broom 丛枝病

Y

yellowing 黄化

Z

zoosporangium 游动孢子囊
zoospore 游动孢子
zygospore 接合孢子
zygote 接合子

附录V 重要的植物病原物属名
Appendix V Genus Names of Important Plant Pathogens

A

Achlya 绵霉属

Agrobacterium 土壤杆菌属

Albugo 白锈属

Alternaria 链格孢属

Anguina 粒线虫属

Aphelenchoides 拟滑刃线虫属

Armillaria 密环菌属

Ascochyta 壳二孢属

Aspergillus 曲霉属

B

Bacillus 芽孢杆菌属

Bipolaris 平脐蠕孢属

Blumeria 布氏白粉菌属

Botrytis 葡萄孢属

Botryosphaeria 葡萄座腔菌属

Bremia 盘梗霉属

Burkholderias 布克氏菌属

Bursaphelenchus 伞滑刃线虫属

C

Ceratocystis 长喙壳属

Cercospora 尾孢属

Chlorocypha 绿杯盘菌属

Ciboria 杯盘菌属

Cladosporium 芽枝霉属

Clavibacter 棒形杆菌属

Claviceps 麦角菌属

Cochliobolus 旋孢腔菌属

Coleosporium 鞘锈菌属

Colletotrichum 炭疽菌属

Coryneum 棒盘孢属

Cronartium 柱锈菌属

Cryphonectria 隐丛赤壳属

Curvularia 弯孢属

Curtobacterium 短小杆菌属

Cucumovirus 黄瓜花叶病毒属

Cuscuta 菟丝子属

Cylindrosporium 柱盘孢属

Cytospora 壳囊孢属

D

Daldinia 轮层炭壳属

Diaporthe 间座壳属

Diplodia 色二孢属

Ditylenchus 茎线虫属

Dothiorella 小穴壳属

E

Erwinia 欧文氏菌属

Erysiphe 白粉菌属

Exobasidium 外担子菌属

Exserohilum 凸脐蠕孢属

F

Fomes 层孔菌属

Fomitopsis 拟层孔菌属

Furovirus 真菌传杆状病毒属

Fusarium 镰孢属

Fusicoccum 壳梭孢属

G

Ganoderma 灵芝属

Gibberella 赤霉属

Glomerella 小丛壳属

Gnomonia 日规壳属

Guignardia 球座菌属

Gymnosporangium 胶锈菌属

H

Heterodera 异皮线虫属

I

Ilarvirus 等轴不稳环斑病毒属

L

Lachnellula 小毛盘菌属

Leptothyrium 细盾霉属

Leptosphaeria 小球腔菌属

Leptostroma 斑壳霉属

Liberobacter 韧皮部杆菌属

Longidorus 长针线虫属

Lophodermium 散斑壳属

Loranthus 桑寄生属

Luteovirus 黄症病毒属

M

Macrophoma 大茎点菌属

Macrophomina 球壳孢属

Marssonina 盘二孢属

Melampsora 栅锈菌属

Melanconis 黑盘壳属
Meliola 小煤炱属
Meloidogyne 根结线虫属
Microsphaera 叉丝壳属
Monilia 丛梗孢属
Monochaetia 盘单毛孢属
Mucor 毛霉属
Mycosphaerella 球腔菌属

N

Nectria 丛赤壳属
Neovossia 尾孢黑粉菌属
Nepovirus 线虫传多面体病毒属
Nyssopsora 花孢锈菌属

O

Oidium 粉孢属
Orobanche 列当属

P

Pantoea 泛菌属
Penicillium 青霉属
Peronosclerospora 霜指霉属
Peronospora 霜霉属
Pestalotiopsis 拟盘多毛孢属
Phakopsora 层锈菌属
Phellinus 木层孔菌属
Phoma 茎点霉属
Phomopsis 拟茎点霉属
Phragmidium 多胞锈菌属
Phyllachora 黑痣菌属
Phyllactinia 球针壳属
Phyllosticta 叶点霉属
Physoderma 节壶菌属
Phytophthora 疫霉属
Phytoplasma 植原体属
Plasmodiophora 根肿菌属
Plasmopara 单轴霉属
Podosphaera 叉丝单囊壳属
Polystigma 疔座菌属
Potyvirus 马铃薯 Y 病毒属
Pratylenchus 短体线虫属
Pseudomonas 假单胞杆菌属
Pseudoperonospora 假霜霉属
Puccinia 柄锈菌属
Pucciniastrum 膨痂锈菌属
Pyricularia 梨孢属
Pythium 腐霉属

R

Radopholus 穿孔线虫属
Ralstonia 劳尔氏菌属
Ramularia 柱隔孢属
Rhizoctonia 丝核菌属
Rhizopus 黑根霉属
Rhytisma 斑痣盘菌属
Rosellinia 座坚壳属

S

Sawadaea 叉钩丝壳属
Schizophyllum 裂褶菌属
Sclerotinia 核盘菌属
Sclerospora 指梗霉属
Sclerotium 小核菌属
Septobasidium 隔担子菌属
Septoria 壳针孢属
Sobemovirus 南方菜豆花叶病毒属
Sphaceloma 痂圆孢属
Sphacelotheca 轴黑粉菌属
Sphaerotheca 单丝壳属
Spilocaea 环痕孢属
Spiroplasma 螺原体属
Streptomyces 链丝菌属
Synchytrium 集壶菌属

T

Taphrina 外囊菌属
Tilletia 腥黑粉菌属
Tobamovirus 烟草花叶病毒属
Trametes 栓菌属
Trichoderma 木霉属
Trichothecium 聚端孢属
Trichodorus 毛刺线虫属

U

Uncinula 钩丝壳属
Urocystis 条黑粉菌属
Uromyces 单孢锈菌属
Ustilaginoidea 绿核菌属
Ustilago 黑粉菌属

V

Valsa 黑腐皮壳属
Venturia 黑星孢属
Verticillium 轮枝孢属
Viscum 槲寄生属

X

Xanthomonas 黄单胞杆菌属
Xiphinema 剑线虫属
Xylella 木质部小菌属